大是文化

商用英語
立可貼

上網開店、網拍、揪團購、甚至開公司，
報價、詢價、殺價、訂約收錢，就算菜英文，

複製貼上就搞定

美國東密西根大學語言及國際貿易系雙學位
臺灣科技大學語言中心講師
林奇臻◎著

CONTENTS

序章　如何寫？關鍵在溝通，而非賣弄

Part 1 招攬交易

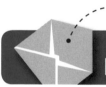

Part 2 詢價、報價、還價

CONTENTS

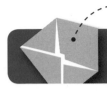

Part 3 簽約成交與履約

Part 4 申請開狀與保險

Part 5 包裝與出貨

CONTENTS

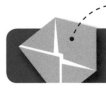

Part 6 運輸

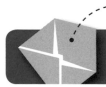

Part 7 付款

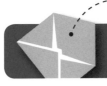

Part 8 申訴與索賠

附錄 快速成交 商用英文書信範例

推薦序一
只要套用範例，
就能輕鬆寫好英文信

文／臺灣科技大學人文社會學院院長　黃國禎

在全球高度競爭的21世紀，國際化已經是所有企業在發展過程中必須面對的問題；因此，商用英語寫作也成為現代人必須具備的能力。然而，對於非母語的學生，寫作往往是學習英語的過程中最大的挑戰；對於教師而言，要在短時間內教會學生撰寫英語書信，更是艱鉅的任務。

很高興有機會拜讀林奇臻老師這本《商用英語立可貼》。這本書將我原本認為十分困難的教學目標，用系統化的方式解決。林老師藉由各種類型的範例作為學生的寫作「鷹架」（輔助工具），引導學生們在很短的時間內完成英語商業書信，並且在教學實務中獲得很好的成果。

對於許多想要進入國際化市場的企業來說，這是一本很實用的內部人員訓練教材；對於想要進入跨國企業的年輕學子或想要進修的上班族，這是一本很有價值的自學工具書。相信本書的出版，必能夠嘉惠許多為學習商業英語寫作所苦的學生及職場人員；同時，書中提供的各種範例，更能幫助教師們在英語課堂中強化學習活動的效果。

推薦序二
先讀過寫作技巧，
忘記了就隨時查

文／新聯翔管理顧問股份有限公司執行長　朱兆港

在寫英文電子郵件的時候，我常常因為郵件開頭用語該怎麼寫感到十分苦惱，不知如何下手。問候或回應對方的話，該從那裡開始？該用什麼情境動詞較恰當？如果對方是外國人（肯定絕大部分是），雙方有著不同的文化、不同的習慣、不同的語言特性，如果一不留意用字，就會鑄成大錯，或者鬧出笑話。

　　奇臻老師在課上分享的這些英語寫作用語，經過整理後編纂成這本《商用英語立可貼》，造福眾人、令人佩服。這本商用英語寫作書籍的問世讓我如獲至寶，建議各位可以先把內容閱讀一遍，試著將自己用得到的內容燒錄在自己的腦海中，永遠不要抹滅。至於，記不起來的，就把它當成工具書，需要時隨時翻閱，這本書一定會成為我的最佳工具和夥伴，相信它不僅會成為我的福音，也會是每一位讀者的最佳良伴。

前言

複製、貼上，30分鐘就寫完，馬上和老外做生意

在這個時代，喜歡英文最好，討厭英文就糟糕了，因為英文還是必須學習的重要語言。在臺灣各大公司，例如科技大廠台積電，他們有50％的產品都行銷到北美地區，許多跨國企業與高科技的人才，都有相當的英語文能力，而我們都得為邁入下一波全球競爭而努力。

我在臺灣念書時，畢業成績並非十分優秀，當年沒有考上最好的國立大學。打工期間，因升學壓力，某天趁著夜深人靜，拿著一本當年專科學校蔡成立老師編寫的商用英文教科書，坐在電腦桌前，打開 Word 檔，一面照著書信範例，一面跟著打字，寫了一封英文詢問信，就寄到美國去了。

很難想像，拿到郵局後的英文信件，會寄到了距離臺灣12,000公里遠的美國大學註冊組，寫信的內容是為了詢問入學資訊。一封封英文書信往來，開始了我大學及研究所的留學生涯。

身為基督徒的我，相信聖經的描述，如經上所記：「神為愛他的人所預備的是眼睛未曾看見，耳朵未曾聽見，人心也未曾想到的（林前2：9）。」往後，不論是申請獎學金、住宿、網路購物，連畢業後的第一份工作——面板公司的總經理秘書，也是藉著努力和信念，靠英文書信應徵而來。

所以，商用英文書信不僅用於求學、生活，更是職場上的主流，

它並不會因為紙本信件或 E-mail 有所不同，也不會因為 Line、Memo 的流行，或手機、電話的快速而取代之。不論是經理人、採購人員、研發人員、業務人員、秘書人員、人資、社會新鮮人或在校生等，每個人都要學會輕鬆的寫商用英文書信。

特別是現在有些青年人創業，如網路拍賣等，經常要到國外批貨、跟外國人做生意。貿易上和交易，都需要具備撰寫英文書信的能力。

於是，在大是文化總經理陳絜吾的構想之下，有機會再開始著筆適合學生，也適合社會新鮮人或職場老手等的商用英文書信自學書。全書主題依整體貿易流程編排章節，讓中英文商用書信範例更簡單易懂，加上內文常用佳句以及單字或關鍵詞句，希望更適合讀者們自學。

在臺灣科技大學教授商用英文書信課程的同時，我在班上也成立了專題小組，讓幾乎沒有寫過商用英文書信的同學，嘗試著在課堂中閱讀範例，經過教授書信結構及自行閱讀後，他們能輕鬆的完成主題和一篇英文信。

另外，也訪談幾位目前在社會商業界、學術界頗具成就，受到大眾肯定的人士們，希望能將他們的個人經驗和讀者分享。

在此，我要致上謝意，感激臺灣科技大學人文社會學院院長黃國禎講座教授、臺灣科技大學應用外文系洪紹挺副教授、語言中心黃淑嬌主任及同仁們的指導與協助。也十分感謝新聯翔管理顧問股份有限公司朱兆港執行長，與產投計畫美語團隊的支持，感謝專家教授們在百忙之中不吝撥出寶貴時間，來和我討論商用英文書信。

感謝專題小組同學（臺灣科技大學）協助整理資料和分享心得：

• 材料系四年級范齡云同學：

很開心能參加這次專題，能把上課學的東西應用於生活上，30分鐘絕對能完成一篇商用信件。信中的英文及文法都是很基礎的，不需要很高的英文程度也能輕鬆完成，我覺得可以讓上班族不用花太多時間構想商用信件，也可以花更多時間在職場上。

• **營建系四年級王聖凱同學：**

很高興有機會參與專題寫作，我覺得用替換關鍵詞句的寫作方法很棒，以前練習的方法都是直接憑空思考，會較吃力且費時，而且文法上也較常出錯。使用這種方式不但讓寫作更容易，也較省時。

• **工管二年級黃怡晶同學：**

我有了新的機會嘗試，從閱讀者的角度轉換到寫信者的角度去寫，閱讀範例和佳句，能快速的寫出我自己想要的書信！

• **電機系二年級林至遠同學：**

這次的專題寫作計畫讓我對商用英文書信的整合和應用更上一層樓了！藉由老師提供的範例和常用佳句，也能建立對書信的架構，未來若有機會使用，肯定能更適當並充分利用。

• **電子系四年級洪紹唐同學：**

在參與專題寫作的過程中，這種替換單詞和運用常用佳句的方式，讓我在文章架構和文法上都沒有遇到太大的障礙。

• **管理學士班李韋澔同學：**

用這種方式寫商用英文書信，就沒有太大的難度，雖然商業上的交易方式是平常不曾接觸到的，但將來若實際應用，利用範例和佳句肯定可以更有效率和展現專業度。

最後，感謝大是文化出版社、家人、朋友和師長同學們的支持和鼓勵。期望這本書能在職場中帶給大家極高的工作效率，讓大家輕鬆成為商用英文書信達人，在短時間內搞定商用英文書信，搞定生意。

序章

如何寫？
關鍵在溝通，而非賣弄

網路的發達，改變了商業環境，其中最大的變化就是交易無國界，也無資源的差異。你可以隨時上網找你想賣的東西，也可以販賣別人想要的東西。因此，越來越多年輕人跑來問我，老師，我在美國網頁上看到一個好炫的產品想引進台灣賣，但我英文不好、不會寫信怎麼辦？

我常說，商用英文不是真正的英文，而是一種溝通工具。寫作時你不該賣弄你的英文能力，而是用對方最容易理解的用字和語彙，在最短的篇幅內傳達你想表達的事。

例如，寫信時，你必須考量對方和你的關係、文化背景、對方公司的專業職務等，來思考信件中使用的文體，也要盡量避免使用過時的片語，注意遣詞用字等等。

接著，就讓我們看看寫一封英文信應該注意哪些步驟與環節吧。

成功的商用書信寫作

第一，落筆前先想，寫這封信要達成什麼目標？

　　與對方聯絡、洽談商務時，要先擬定一個清楚的目標（例如是為了報價、詢價或申訴等），釐清寫這封英文信的目的。 接著，要規劃一下要花多久時間寫這封信，並想想看要寫什麼樣的內容，以及要寫多少字，最後把你想到的寫進信裡。

第二，這封信你要寫給誰，該用什麼樣的語氣？

　　當我們寫信向廠商詢問報價或聯絡廠商時，要先了解閱讀信件的人可能會是誰，信中只能寫相關的資訊。

　　當然，用語也要保持客氣和禮貌。同時也要注意，信中的用語要使用對方適應的風格和筆調（要思考一下對方與你的關係為何），同時最重要的是，先了解對方或對方公司的專業職務，也就是了解對方公司內部、外部的階層。

　　接著，必須考量對方對信中主題有多麼熟悉（例如想想對方的文化背景、年齡、性別，也要避免信中的用詞帶有性別歧視的意涵）。

第三，選擇正確的語言

　　聯絡的內容要言簡意賅，不要過度冗長。信中每一句約15到20個字，每段約7行到8行，甚至只有3到5行也無妨。

　　建議使用簡單的日常英語，避免使用專業行話、縮寫和複雜的短語。同時要注意不要重複使用單字和片語。使用片語的時候，也須注意使用正確的片語開頭和結尾。此外，文法、拼字與標點符號的使用，更得多加留意。

最後，評估、修訂和寄出

當內容準備就緒之後，記得留點時間重新檢視信件格式、語言是否有誤？內文的風格是否恰當？最好可以把信件大聲唸出來，確認一下文字的流暢度。最後確認一下文法、標點符號和拼字有沒有錯誤。

總結

落筆前先規畫大綱 → 考量讀者 →
把資訊結構化 → 選擇正確的語言 →
打草稿、評估和修正 → 寄出

寫信時先想：

為什麼要寫這封信？為什麼不用電話聯絡對方？

因為：

- 人們需要書寫紀錄。因為訊息內容往往很複雜。
- 將內容寫下來可以幫助釐清思緒。
- 不需要即時做出回應。
- 每個人都可以獲得相同的資訊。
- 可以提升個人的權威。
- 能夠塑造企業形象。

1. 寫作文體分三類，根據對象決定

英文書信共有三種主要的寫作形式，即正式、標準、非正式。

大部分的商用書信使用的是標準形式，然後再根據你的聯絡對象，加入其他兩種形式的元素。

至於哪一種書信寫作的風格最好？其實還是要根據你的聯絡對象來決定。要注意的是，內容可別寫得太正式，也不能寫得太過隨便，必須拿捏恰當，保持你的專業程度。

正式商業書信寫作文體

正式的商業書信寫作文體種類很多，例如法律上的契約必須使用正式的語言，也要避免使用過時的片語。信件內容必須用詳細且縝密的說明，利用客觀且間接委婉的語調，非常有條理的闡述內容。在文法與標點符號的運用上，也必須完美無缺。

標準商業書信

在標準的商業書信中會使用專業用語，其中再搭配較淺顯易懂的短語，也要注意不可以使用太多的正式語彙。書寫風格上，可以採用稍微個人和直接的風格，內容要有條理，當然在文法與標點符號的運用上也不可以有錯。

非正式電子郵件

　　至於非正式的電子郵件，可以採用非正式的語言，不妨透過會話的形式來表達，內容也可以搭配一些日常用語。在信件的風格上，不妨添加一些個人風格，或許可以比較貼近讀者，詞彙用語的使用上可以不那麼拘謹。基本上，信件內容可以稍微接受不是那麼完美的文法與標點符號運用。

寫作後檢查表

　　當你寫完一封英文書信後，可以利用下方的檢查表，檢查一下信件的內容是不是都具備了下列的條件：

- 目的明確　　　　　　　　　　☐
- 良好的內容組織　　　　　　　☐
- 專業的格式　　　　　　　　　☐
- 清楚且簡要的語言　　　　　　☐
- 正確的寫作形式　　　　　　　☐
- 恰當的陳述和圖表　　　　　　☐
- 已回答信中所有問題　　　　　☐
- 積極的結尾　　　　　　　　　☐
- 正確的文法　　　　　　　　　☐
- 拼字及標點符號運用是否正確　☐

簡短格式書信

　　若是格式比較簡短的書信，須注意信件內容要前後一致，而且這種簡短格式千萬不可以用在正式的書信。想要寫好一封簡短格式的商用英文書信，必須清楚表達你想要傳遞給對方的訊息，並且在主題上，找出能和對方協調的寫作方式和筆調。

注意事項

書信的語言

　　寫英文書信時，信件的開頭可以運用不同的句型與佳句，句子要有重點、簡潔有力，使用淺顯易懂的詞語。最好可以用關鍵詞句做標語、開始及結束段落。不要使用粗俗的字眼和贅詞，以及不適當的省略字句。標題要確保清楚簡潔。也要謹記不可重複使用字詞或短語，好讓信件的內容更加豐富。

比較容易讓讀者接受的書信

　　要怎麼讓對方在閱讀你的信件時，可以較輕鬆、沒有壓力呢？以下有幾個小技巧提供大家參考。

　　首先，要確認文字之間的行距大小一樣，而且要保留足夠的空間。信件頁邊留白的部分，不可以少於2.5公分。寫信時也要避免段落冗長，記得要利用粗體和底線的大標及小標，標示出重要的地方，這樣可以讓對方更快的抓住重點。

還有，當信件內容使用到數字的時候，要注意格式得前後一致，不可以一下子使用「12345」，一下子使用「一二三四五」，以避免對方閱讀信件時混淆。

語調

在寫英文書信時，不妨問問自己：「我想要寫什麼樣的內容？」、「我想藉由這封信達到何種目的？」在寫信時，記得任何時候都要保持有禮且專業。當信中會提到一些負面的內容時，可以在開頭與結尾使用正向的文字包裝負面的內容。

使用正確的詞語傳達你的意見：

建議使用的片語：

• 強烈的

We strongly recommend that......　　我們強烈建議……

We are absolutely convinced that......　　我們絕對相信……

• 中立的

We recommend that......　　我們建議……

We expect that......　　我們期望……

• 委婉的

One option may be to......　　有一種選擇是……

We could consider......　　我們會考慮……

2. 從架構大綱到三種敘事方法

　　某些商業上的書信寫作，也許會在沒有太多時間可以做事先準備的狀況下完成。不過，比較複雜的資訊都需要事先計畫與整理。這裡也提供了一些實際的技巧供大家使用。

如何制定計畫

書信綱要計畫

　　①首先，要寫下你寫這封信的目的、目標，②接著按照你的目標思考可行的內容（或是替代方案），③再依據優先順序選擇你需要的資訊。④當一切都準備就緒之後，還要準備一個扼要、標題來概括信件的重點，好讓對方可以一目了然，了解信件的大致內容。⑤當你想好所有細節之後，就可以開始撰寫（起承轉合）英文書信了。

編制、組織資訊的方法

　　有許多不同的方式可以組織信件資訊。以下是一些常見的技巧。

金字塔型

先概括主題，再逐步提出各項細節。這個方法最常用於報告及宣傳內容（由大至小）。

Our customers expect the best in terms of service quality.
So, we prioritize reliability, availability, and safety in our systems.

我們的客戶期待得到最好的服務品質。
所以，我們優先考慮系統的可靠性、可用性和安全性。

我們可以將金字塔「反轉過來」，從特定的事件延伸到大主題。能有效的應付需表明特定規則的情況（由小至大）。

Last week, a member of staff parked in the CEO's parking space.
May I remind staff that there are spaces reserved for senior management.
Any further incidents of this nature will be dealt with severely.

上個禮拜有一位員工將車停在執行長的停車位。
容我提醒員工們，有些車位是留給高階主管的。
再有類似的事件發生，將會嚴厲處理。

商用英語立可貼

年表法

按時序敘述事件，按時間一格一格的敘述。

The negotiation went well underline{yesterday}. We managed to agree to quantities and price. We spent a lot of time underline{today} sorting out terms of payment and a timetable for delivery. We should be in a position to sign the contract underline{tomorrow}.

昨天的協商很順利。我們雙方同意了數量和價格。今天我們花了很多時間整理出付款的項目以及運送的時程表。我們預計明天就可以簽署合約。

從問題伸延到解決方法——先敘述發生的問題，再根據問題回答（自問自答）。

Several users have reported that their new passwords no longer allow them to access to the Intranet. The problem was caused by a software error. Please enter and confirm a new password with a minimum of seven, not six characters.

很多使用者反映無法使用新密碼登入內部網路。這是一個軟體發生錯誤所造成的。請輸入並確認 7 位字元的新密碼，而不是 6 位字元。

按程序排列——按照程序清楚的排列出來（一個口令一個動作）。

To activate Voicemail from London： *1. Dial 134.* *2. Listen to the information message.* *3. Activate Voicemail by pressing 1.* 欲從倫敦啟用 *Voicemail*： 1. 撥 134。 2. 聆聽資訊。 3. 輸入 1 以啟用 *Voicemail*。	*Make it Clear* *This instructional format lists three points making it clear and easy to read.* 清晰說明： 這個範例簡易清晰地列出每一項重點，以便對方閱讀。

典型敘事法

敘事分為開頭、中段、結尾，很多信件都會使用這種簡易清晰的三段落格式（啟承合）。

I'm sorry that I wasn't able to make the meeting last Friday. 我很抱歉上星期五無法出席開會。 （I 概述）
Unfortunately, we had a major problem at our plant in Berlin, which I had to solve. 不幸地，當時我們在柏林的工廠出了個重大問題，我必須前往處理。 （II 延伸）

Fortunately, I managed to sort things out. I will be in Paris next week as planned and look forward to seeing you in order to finalize the outsourcing issue.

幸好，我把事情解決了。下星期我將會依照原計劃到巴黎，很期待見到您以敲定外包事項。

（*III* 結論）

3. 架構總整理，禮貌別忘記

一封商用英文書信中，整體內容應包括如下：

（1）Letter head（信頭）

（2）Date（日期）

（3）Ref. No.（參考案號）

（4）Address（收信者地址）

（5）Attention（重要收信人）

（6）Salutation（敬稱）

（7）Subject（主旨）

（8）Body of Letter（內文）

（9）Closing Sentence（結尾佳句）

（10）Complimentary Close（客套結語）

（11）Signature（寫信者的親筆簽名、打字名及職稱）

（12）ID Initials（識別記號）

（13）Encl.（附件）

（14）c.c. or cc（副本抄送單位）

（15）Postscript（附記）

• Letter head（信頭）：

英文信紙上端印有發信公司訊息，

商用英語立可貼

例如：ⓐ 公司名稱及地址（中文及英文）與信箱號碼。ⓑ 電話及傳真或電子郵件信箱。ⓒ 公司商標等相關資訊。

- Date（日期）：

美式：June 18, 2016

英式：18th June 2016（2016年6月18日）

- Reference Number（參考案號）：

收發信雙方便於檔案整理用。

例如：Our Ref. 12345.（我們的參考案號12345）. Your Ref. Our Fax（我們的傳真）.

- Address（收信者地址）、Inside Address（信內地址）：

收信人姓名及地址，例如：（按地址排列）

Mr. George Wu	收信人名稱
NTUST Software Co. Ltd.	公司名稱
No. 195, Beacon St.	號碼，路名／街名
Boston, MA 0321,	鄉鎮縣市，省份（州）
U.S.A.	國家

- Attention 指定收信人（收信部門）：

例：Attention：Sales Department（收信人：銷售部門）

例：Attention of Mr. George Hwang（收信人：黃喬治先生）

例：Att：Mr. Michael Chou（收信人：周麥可先生）

- Salutation（敬稱）：

除了 Dr.、Mr.、Mrs.、Ms. 外，敬稱不可用簡寫。頭銜一律採用英文大寫。

例：Dear Mr. Hwang,

例：Dear Dr. Hwang,

例：Dear Miss,

例：Dear Gentlemen,

例：Dear Ladies,

例：Dear Sir,

- Subject（主旨）：

商業信件為了方便閱讀，也可在 Subject 字下加一橫線，使對方可以清楚了解信件的主旨。

- Body of Letter（內文）：

格式：

ⓐ Indent Form（縮退式）：每段首行須後退五個字。

ⓑ Block Form（採齊頭式）時：各行均由 Marginal Line（邊緣）左側開始書寫。ⓐ、ⓑ 為常用書寫格式型態。

ⓒ 段落間間隔為 1 行。

ⓓ 段落中單行間距。

ⓔ 插入表格，邊緣起左邊留 8 字，右邊留 5 字，上一段落至表格間留 2 行。

內文結構：

商用英語立可貼

內容結構可約略分為開頭、中間和結尾。

Ⓐ開頭：簡短陳述寫信理由，強調與對方之相關關係與利益。

Ⓑ中間：說明自己的優點及期望，所提之事使對方信任並說服對方依自己做法行事；或告知其分析事情的優劣結果。

Ⓒ結尾：根據信件內容而變化，期待的對方回應有三種：

① 期待對方同意：可以在結尾寫出要對方回答的問句或請教語，寫上期待對方回覆之文句。

② 期待對方或祈求對方見諒。

③ 期待對方「說不」：內容引導對方主動說不。在信中部分先做分析說明，再於結尾部分建議對方改變預定的做法，用語須柔中帶剛，說服對方。

- Closing Sentence -The closing sentence is the last sentence in a paragraph.（結尾佳句）：

  結尾請用句號「．」。

  I look forward to hearing from you soon.
  （期待您的即時回應）

  Thank you for your time and attention.
  （謝謝您的時間與關注）

  Thanks very much for your help.
  （非常謝謝您的協助）

  Thank you in advance for your kind attention.
  （先謝謝您貼心的關懷）

  We hope to receive your reply.（期待您的回覆）

  Please let me know.（請讓我知道）

- Complimentary Close（客套結語）：

  禮貌必備用語，務必加上逗號「,」。

  Sincerely yours,（您誠摯的）

  Yours sincerely,（您誠摯的）

  Yours faithfully,（您忠實的）

  Cordially,（誠摯的）

  Best regards,（最好的問候）

  Thanks and best regards,（感謝並致上最誠摯的問候）

  Kindest regards,（最貼心的問候）

  Yours truly,（您真實的）

  Very truly yours,（您最真誠的）

  Best Wishes,（送上我最佳的祝福）

- Signature（簽名）：

  寫信者務必親筆簽名，其它內容可打字。

  公司名稱（打字）負責人簽名（簽字，Email中則無親筆簽名）。

  寫信者職稱（打字）及所屬部門。

  例如：

  Chi-Jen Lin,

  English Instructor, Language Center,

  National Taiwan University of Science and Technology

  No. 43, Sec. 4, Keelung Rd., Taipei 106, Taiwan

商用英語立可貼

- Identification Initials（識別記號）：
  請以「簽名者、打字者」順序寫在信件之左下方，位於簽名之下1行並由邊線開始書寫。

- Enclosure Remarks（附件）：
  表隨同信件之文件。用Enclosures或Encls. 來表示，例如：Encls. 3（信件中有3個附件）或 Encl.：catalog 1（信件中有含1目錄）。

- Carbon Copy（副本抄送單位）：
  即我們常用的c.c.和C.C.。

- Postscript（附記）：
  常用 P.S.表示正文所寫之附記，結尾後由簽字者簽上姓名之第一個字母以示負責。

4. 常用短語及單字，直接套用就對了

Staring 開頭

- With reference to your letter of 16 April, ...
  根據您4月16日所寄的信件……

- Regarding our meeting last week, ...
  關於我們上星期的會議……

- Thank you for your letter of...... (date)
  謝謝您……（日期）的來信

- In reply to your fax, ...
  回覆您的傳真……

Reason for writing 寫信的理由

- We are writing to
  我們寫這封信是為了……

- I'm just writing to
  我寫這封信只是為了……
  （＋request 請求／confirm 確認／inform 告知）。

- Just a short note to
  只是想簡短的提醒您……
  （＋ask 詢問／check 核對）。

商用英語立可貼

Giving good news 告知好消息

- We are delighted to inform you that......
  我們很高興的通知您……
- You will be pleased to hear that......
  相信您會很樂意聽到……
- You will be happy to learn that......
  相信您會很高興得知……

Giving bad news 告知壞消息

- We regret to inform you that......
  非常抱歉通知您……
- I am afraid that....../Unfortunately......
  恐怕……／非常不幸的……
- I am sorry but......
  我很遺憾要告訴您……

Making a request 請求

- We would appreciate it if you could......
  如果您能……我們會非常感謝
- I would be grateful if you could......
  如果您能……我將會很感激
- Could you......？
  不知道您能否……？

Offering help 提供協助

- If you wish, we would be happy to......
  如果您願意，我們將會非常樂意……

- Would you like me to...... ？／Shall I...... ？
  需要我為您………嗎？／我能為您……嗎？

- Do you want me to...... ？
  您想要我為您……嗎？

Apologizing 道歉

- We must apologize for（not）......／We deeply regret......
  我們要為……道歉／我們對於……深感遺憾

- I do apologize for......（any inconvenience caused.）
  我們為……（造成任何的不便）感到抱歉

- I'm really sorry for／about......
  我對於……感到非常抱歉

Enclosing documents 附加檔案

- We are enclosing......／We enclose......
  我們附上……（的檔案）

- Please find enclosed......
  請查看附檔裡的……

- I am enclosing......／I have enclosed......
  我附上……（的檔案）／我已附上……（的檔案）

Closing remarks 結尾語

- Do not hesitate to contact us again if you need further assistance.
  如果您需要進一步的協助，請馬上聯繫我們。
- If you have any further questions, please contact me.
  如果您有任何其他的問題，請聯絡我。
- Let me know if you need any more help.
  如果您需要任何協助，請讓我知道。
- Don't forget! -Thank you for your help.
  不要忘了！──謝謝你的協助。

Positive future reference 建立未來的正面關係

- We look forward to meeting／seeing you next week.
  我們很期待再跟您見面／與您下禮拜見面。
- We look forward to hearing from you.
  我們很期待再收到您的消息。

書信短語與單字

Nouns 名詞

- assistance 幫助；援助
- correspondence 信件；通信
- expression 表達；表示

- inconvenience 不便之處；麻煩

Verbs 動詞

- apologize 道歉；認錯
- contact 聯繫；聯絡
- enclose 附上；封入
- hesitate 猶豫；躊躇
- inform 告知；通知
- look forward to 期待
- regret 遺憾；懊悔

Adjectives 形容詞

- acceptable 可接受的；可忍受的
- conversational 會話的；健談的
- direct 直接的
- elaborate 詳盡的；精巧的
- further 進一步的；另外的
- grateful 感激的；感謝的
- indirect 間接的
- old-fashioned 舊式的；過時的

Adverbs 副詞

- deeply 強烈地；深刻地
- unfortunately 不幸地；遺憾地

5. 國際貿易實務流程，
做生意的眉角在這裡

　　上網購物或開設網站商店，成為這幾年來非常流行的消費與經營模式，想做生意，國貿實務也成了重要的常識。從詢價、報價、簽約到處理客訴等，都是經常遇到的狀況。本書除了介紹商用書信的寫法外，也介紹大略的貿易流程，讓大家在處理商務上更得心應手。

招攬交易

　　招攬交易，一般可透過出國拜訪、刊登廣告、寄發推銷信及參加展覽等方式來進行市場調查，尋找交易對象，進行交易之提議。

詢價、報價、還價

　　對於感興趣的商品，買方可以進一步向賣方發信詢價，賣方則可根據買方需要的商品數量或其他條件進行報價。但在市場價格競爭激烈的時代，買方通常會同時向多家廠商詢價，經過多方比價及還價後，慎選其中一家或一定的廠商合作。

簽約成交與履約

在貿易實務程序中，稱為接受並確認，也稱是簽定契約。買方在接受報價，再經賣方最後確認，買賣契約即告成立。買賣契約一旦成立，雙方的權利義務也因而確定，賣方須交付貨品，買方則須支付貨款。

申請開狀與保險

以信用狀為付款方式，買方必須向往來銀行申請開立信用狀，開狀銀行受理買方之申請後，將依買方指示簽發信用狀。開狀銀行開立信用狀後，一般會透過其在出口地之往來銀行通知信用狀，通知銀行接到國外開狀銀行開來的信用狀。予以查對後，通知信用狀受益人前來領取。

另外，FOB（Free On Board，離岸價，表示賣方將貨物交到出口港海洋輪船上，責任即告解除，此後的費用與風險均由買方負擔」），CFR（Cost and Freight，根據本條件，賣方應負責找船，洽定船位，並支付海運費，交貨地點與FOB一樣，仍為出口地海洋貨輪上。）貿易條件下，通常會要求買方在申請開狀前先投保貨品運輸保險。

在CIF（Cost、Insurance and Freight，根據本條件，買賣時雙方除了負責CFR條件的義務外，賣方還要負責購買貨物的海上運輸保險，支付保險費）貿易條件下，賣方須向保險公司投保貨品運輸保險，取得保險單做為押匯文件（參考資料：定期航運貨櫃運輸實務教學網站）。

包裝與出貨

賣方若發現貨品包裝不一致，需要延遲交貨期；或買方對包裝的反映，缺貨等事宜，都需盡快在製造過程通知對方。賣方在貨品出貨後，應發給買方有關貨品詳細資料及裝運情況的通知，讓買方能做好籌措資金、付款、進口通關和提貨的準備。

運輸

發貨通知應按契約或信用狀規定的時間發出。賣方可詢問貨運公司運費，到貨時通知買方到達進口地，若不幸發生事故，也必須在第一時間通知對方。買方也可發信詢問貨櫃情況等。

付款

　　在貨品裝船出口後，賣方要遵從信用狀之規定，備齊信用狀規定之匯票、發票、提單、保險單等單據，向往來銀行申請押匯。押匯銀行依指定，將單據寄往開狀銀行或其指定銀行請求付款。買方付款後，開狀銀行即將貨運單據交與買方。

申訴與索賠

　　商務活動中，針對對方索賠函中提出的索賠意願和要求而回覆的申訴處理信函。從樹立良好形象及未來的業務拓展角度而言，無論企業還是個人，在收到索賠函後都應當認真對待、及時處理，以誠懇友好的態度處理問題，有錯認錯，無則嘉勉。通常來說，理賠函包括以下主要條款：1、說明來函目的；2、糾紛處理態度及意見；3、賠償處理意見；4、感謝並期待未來繼續合作。

Part 1
招攬交易

推銷個人或推銷公司產品是產生交易的
第一步。招攬交易，一般可透過寄發推
銷信、刊登廣告等方式進行，尋找交易對
手，進行交易之提議。

書信範例

NTUST
No. 1,
Section 4,
Roosevelt Road,
Da'an District 10617,
Taipei City, Taiwan

September 25th, 2015

Dear Mr. Chen,

Thank you for your attention. We were given your information by our mutual colleague, Mr. Wang. I would like to take this opportunity to introduce ①NTUST Camera Device in the hopes that we may work together in the future.

②NTUST Camera Device was established in 1996 and since then has grown to be the primary supplier of cameras for Taiwan, Japan and and USA. Our ③Taiker Camera is currently the most popular item in Japan and Taiwan. Last year we were awarded an AAA business credit rating by East Asian Trade Board, their highest category of reliability.

We are currently looking to expand our foreign market, and we believe that we rate the best ④cameras by price, features and value. I have enclosed information about the ⑤product lines and our company ⑥history for your reference.

Thank you again for your attention. I hope to hear from you soon.

⑦Best regards,

Kai

Vice president
Marketing Department
NTUST Camera Device

Encl.1

中文翻譯

臺北市 10617 大安區羅斯福路 4 段 1 號　NTUST

2015 年 11 月 25 日

陳先生您好，

感謝您的關注。我們收到我們彼此認識的同行王先生給予的聯絡資訊，想透過這次機會向您介紹我們公司的 NTUST 照相機，並希望未來有機會和您合作。

NTUST 攝影器材公司自 1996 年成立以來成長茁壯，在臺灣、日本和美國等地，現已成為主要相機供應商。我們的 Taiker 照相機目前是臺灣和日本最受歡迎的產品。去年，我們公司榮獲東亞貿易局頒予 AAA 企業信用等級，此為該局給予對可靠性評價的最高評級。

我們正在努力拓展國際市場，而我們認為本公司的照相機在價格、特色和價值上極有競爭力。我已附上有關生產線以及公司沿革的資料給您參考。

再次感謝您的關注。希望很快能收到您的回覆。

最誠摯的問候，

凱

副總裁
行銷部
NTUST 攝影器材公司

附件 1

✓ 內文佳句

1. I would like to take this opportunity to introduce ① _____ in the hopes that we may work together in the future.

   想透過這次機會向您介紹我們公司的 _____，並希望未來有機會和您合作。

2. ② _____ was established in 1996.

   _____ 公司自1996年成立。

3. Our ③ _____ is currently the most popular item in Japan and Taiwan.

   我們的 _____ 目前是臺灣和日本最受歡迎的產品。

4. We are currently looking to expand our foreign market, and we believe that we rate the best ④ _____ by price, features and value.

   我們正在努力拓展國際市場，而我們認為本公司的 _____ 在價格、特色和價值上極有競爭力。

5. I have enclosed information about the ⑤product lines and our company ⑥_____ for your reference.

   我已附上有關 _____ 以及公司 _____ 的資料給您參考。

6. Thank you again for your attention.

   再次感謝您的關注.

7. ⑦Best regards,

   最誠摯的問候，

▶ 關鍵詞句

1	opportunity 機會，良機，時機	**n.** 名詞
2	mutual colleague 彼此認識的同行	**n.** 名詞
3	establish 建立，設立，創辦	**v.** 動詞
4	primary 主要的，首要的，基本的	**adj.** 形容詞
5	supplier 供應者，供應商	**n.** 名詞
6	the most popular 最受歡迎的	**adj.** 形容詞（最高級）
7	category 種類，類別	**n.** 名詞
8	feature 特色	**n.** 名詞
9	reliability 信任度	**n.** 名詞
10	expand 擴充，發展	**v.** 動詞
11	product line 生產線	**n.** 名詞
12	reference 參考，查閱	**n.** 名詞
13	attention 關注，注意	**n.** 名詞
14	vice president 副主席，副總裁	**n.** 名詞
15	marketing 行銷	**n.** 名詞

▶ 同義詞

1 introduce＝present 介紹

2 establish＝set up 建立

3 popular＝famous 聞名的／受歡迎的

4 award＝grant 授予

5 expand＝extend 擴大

National Taiwan University of
Science and Technology
No. 43, Section 4,
Keelung Road 106,
Taipei, Taiwan

May 8, 2016

Dear Mr. Yang,

I am the student chairman of the ① New Technology Committee at the ② National Taiwan University of Science and Technology, and I am interested in learning more about this ③ new product which is the world's smallest iPhone card reader, Gigastone CR-8600. Please send me the ④ product catalog and any ⑤ related information at your earliest convenience. If you have any other related products, I would appreciate receiving that information as well.

Thank you for your information and attention.

⑥ Yours truly,

Bruce Chang

National Taiwan University of Science and Technology
EE Department

中文翻譯

臺灣臺北市大安區基隆路4段43號　臺灣科技大學

2016年5月8日

親愛的楊先生，

我是臺灣科技大學 新科技委員會的學生主席，有興趣想認識更多這項新產品，最新蘋果手機專用讀卡器（Gigastone CR-8600）。希望您方便的時候，盡快傳送產品型錄跟相關資訊給我。如果您還有其他相關產品，我也會很樂意接受這些資訊。

感謝您的資訊和關注。

您真誠的，

張布魯斯

臺灣科技大學
電子工程學院

商用英語立可貼

1 I am the student chairman of the ①_____ at the ②_____
_____ .

我是 _____ 的學生主席。

2 I am interested in learning more about this ③_____.
有興趣想認識更多這項 _____。

3 Please send me the ④_____ and any ⑤_____ at
your earliest convenience.
希望您方便的時候，盡快傳送 _____和_____給我。

4 If you have any other related products, I would appreciate receiving
that information as well.
如果您還有其他相關產品，我也會很樂意接受這些資訊。

5 Thank you for your information and attention.
感謝您的資訊和關注。

6 ⑥Yours truly,
您真誠的

1 ...... learn more about ...... 認識更多有關……
I am interested in learning more about this _____ which is
the world's smallest iPhone card reader, Gigastone CR-8600.

有興趣想認識更多這項 _____，最新蘋果手機專用讀卡器
（Gigastone CR-8600）。

▶ 關鍵詞句

1 committee 委員會	**n.** 名詞	
2 chairman 主席	**n.** 名詞	
3 national 國立	**adj.** 形容詞	
4 technology 科技	**n.** 名詞	
5 university 大學	**n.** 名詞	
6 science 科學	**n.** 名詞	
7 interested 感興趣	**adj.** 形容詞	
8 product 產品	**n.** 名詞	
9 card reader 讀卡器	**n.** 名詞	
10 catalog 型錄	**n.** 名詞	
11 convenience 方便	**n.** 名詞	
12 commercial 商用的	**adj.** 形容詞	
13 receive 接受	**v.** 動詞	
14 as well 也	**adv.** 副詞	
15 department 學院、部門	**n.** 名詞	

▶ 同義詞

1 product＝commodity 商品

2 send＝deliver 發送

3 smallest＝minimal 最小

4 related＝relevant 相關

5 receive＝get 取得

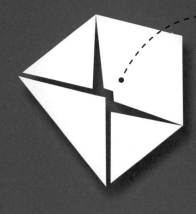

Part 2
詢價、報價、
還價

對於感興趣的商品,買方可以進一步向賣
方發信詢價,賣方則可根據買方需要的商
品數量或其他條件進行報價。但在市場價
格競爭激烈的時代,買方通常會同時向多
家廠商詢價,經過多方比價及還價後,慎
選其中一家或一定的廠商合作。

書信範例

NTUST JACK RESTAURANT
No. 5, Section 4,
Roosevelt Road,
Da'an District 10617,
Taipei City, Taiwan

October 2, 2015

Dear Mr. Tiger,

I am interested in your ①Apple and ②Apple Juice, and would like to receive more information about them, including a ③quotation.

We will request the following products:

1. "Fuji Apple"
2. "Golden Delicious Apple"
3. "Red Delicious Apple"
4. "Cortland Apple"
5. "100% Concentrated Apple Juice"

Please also indicate your earliest delivery date, terms of payment, and discounts for regular purchases. We would appreciate receiving your latest ④catalog as well.

⑤Respectfully,

Jack Lee

NTUST JACK RESTAURANT

中文翻譯

臺北市 10617 大安區羅斯福路 4 段 5 號 NTUST 傑克餐廳

2015 年 10 月 2 日

親愛的泰格先生：

我對於您的蘋果和蘋果汁很感興趣，希望能得到更多有關這些產品的資訊和報價。

我們需要的產品如下：

1.「富士蘋果」
2.「金冠蘋果」
3.「紅元帥蘋果」
4.「科特蘭蘋果」
5.「100% 濃縮蘋果汁」

請表明最早的出貨日期、付款條件以及定期購買的折扣。我們還希望能收到您最新的型錄。

尊敬的，

李傑克

NTUST 傑克餐廳

商用英語立可貼

1. I am interested in your ①_____ and ②_____ ,
   我對於您的 _____ 和 _____ 很感興趣，

2. I would like to receive more information about them, including a ③_____ .
   希望能得到更多有關這些產品的資訊和 _____ 。

3. We will request the following products:
   我們需求的產品如下：

4. We would appreciate receiving your latest ④_____ .
   我們還希望能收到您最新的 _____ 。

5. Please also indicate your earliest delivery date, terms of payment, and discounts for regular purchases.
   請表明最早的出貨日期、付款條件以及定期購買的折扣。

6. ⑤Respectfully,
   尊敬的，

✓ 片語

1. ...... be interested in ...... 很感興趣……
   I am interested in your _____ and _____ .
   我對於您的 _____ 和 _____ 很感興趣。

2. ...... would like to receive ...... 希望能得到……
   We would like to receive more information about them.
   希望能得到更多有關這些產品的資訊。

▶ 關鍵詞句

1	including 包括	**p.p.** 介系詞
2	quotation 報價	**n.** 名詞
3	request 需求，要求	**v.** 動詞
4	following 以下，下列	**adj.** 形容詞
5	delicious 美味的	**adj.** 形容詞
6	concentrated 濃縮的，集中的	**adj.** 形容詞
7	indicate 表明	**v.** 動詞
8	earliest 最早的	**adj.** 形容詞
9	delivery 出貨	**n.** 名詞
10	terms 條件，項目，條款	**n.** 名詞
11	payment 付款	**n.** 名詞
12	regular purchase 定期購買	**n.** 名詞
13	catalog 型錄	**n.** 名詞
14	restaurant 餐廳	**n.** 名詞

▶ 同義詞

1 include = involve 包含

2 quotation = quote（動詞）報價

3 indicate = declare 表明

4 delivery = shipment 運輸

5 latest = current 最新的，最近的

GOOD WINE
No. 502, Xinyi Rd.,
Taipei, Taiwan
(02) 2122-1388

March 30, 2015

Lily Fan
BEAUTY RESTAURANT
No. 8, Song San Rd.,
Taipei, Taiwan

Dear Ms. Fan,

Thank you for your inquiry of ① March 25. I am pleased to provide you with the price list for our ② Nebbiolo wines. The prices quoted here are valid for the next ③ 60 days.

Nebbiolo, 1970, NTD＄6000/bottle
Nebbiolo, 1965, NTD＄6500/bottle
Nebbiolo, 1960, NTD＄7000/bottle

We are happy to offer you ④ a 5% cash discount on payments if you order ⑤ over 20 bottles. Please do not hesitate to contact me if you need any further information.

⑥ Faithfully yours,

Danny Cheng
General Manager
GOOD WINE

臺灣臺北市信義路502號　美酒公司

（02）2122-1388

2015年3月30日

臺灣臺北市松山路8號　美女餐廳
范莉莉

親愛的范小姐：

感謝您在3月25日來信詢價。我很高興向您提供內比奧羅紅酒的價格表。以上所提到的價格，有效期限為接下來的60天。

內比奧羅，1970，新臺幣6000元／瓶
內比奧羅，1965，新臺幣6500元／瓶
內比奧羅，1960，新臺幣7000元／瓶

若您訂購20瓶以上的紅酒，結帳時我們將提供您5%的現金折扣。若您需要任何其他相關資訊，請盡速聯絡我。

您忠誠的，

程丹尼
總經理
美酒公司

✔ 內文佳句

⓵ Thank you for your inquiry of ①＿＿＿＿＿＿ .
感謝您在 ＿＿＿＿＿＿ 來信詢價。

⓶ I am pleased to provide you with the price list for our ②＿＿＿＿ .
我很高興向您提供 ＿＿＿＿＿＿ 的價格表。

⓷ The prices quoted here are valid for the next ③＿＿＿＿ .
以上所提到的價格，有效期限為接下來的 ＿＿＿＿＿＿ 。

⓸ We are happy to offer you ④＿＿＿＿＿＿ cash discount on payments if you order ⑤＿＿＿＿＿ .
若您訂購 ＿＿＿＿＿＿ ，結帳時我們將提供您 ＿＿＿＿ 的現金折扣。

⓹ Please do not hesitate to contact me if you need any further information.
若您需要任何其他相關資訊，請盡速聯絡我。

⓺ ⑥Faithfully yours,
您忠誠的，

▶ 關鍵詞句

1 商品價格表：

商品 Product	標號 Item Number	單價 Unit Price
書 books	IU025	NTD $ 600
手機 cell phones	OP850	NTD $ 8000
電腦 computers	YH320	NTD $ 100000

❷ 價格優惠：

a 5% cash discount on payments if you order over 5 books.

訂購5本書以上享95折

a 10% cash discount on payments if you join our website's

member.

加入網站會員享九折

no shipping cost if you order over 10 cell phones.

訂購10支手機免運費

❸ 結尾敬語：Sincerely 此致、Best regards 最誠摯的問候、

Yours truly 您真誠的

❹ 寄信者職位：

Manager 經理、Assistant 助理、President 總裁

❺ inquiry 查詢，詢價	**n.** 名詞
❻ pleased 高興	**adj.** 形容詞
❼ provide 提供	**v.** 動詞
❽ valid 有效	**adj.** 形容詞
❾ offer 提供	**v.** 動詞
❿ discount 折扣	**n.** 名詞
⓫ payment 付款	**n.** 名詞
⓬ bottle 瓶	**n.** 名詞
⓭ hesitate 遲疑	**v.** 動詞
⓮ contact 聯絡	**v.** 動詞
⓯ further 進一步	**adj.** 形容詞

書信範例

BEAUTY RESTAURANT
No. 8, Song San Rd.,
Taipei, Taiwan
(02) 2202-5948

April 1, 2015

Danny Cheng
GOOD WINE
No. 502, Xinyi Rd.,
Taipei, Taiwan

Dear ①Mr. Cheng,

Thank you for your quotation dated ②March 30 for ③Nebbiolo wines.

We regret to inform you that our company is unable to place an order, because your prices are much higher than those I have been quoted by other dealers. Please let me know if you can reduce your quoted prices.

Thank you for your attention. We look forward to hearing from you.

④Faithfully yours,

Lily Fan

Lily Fan
General Manager
BEAUTY RESTAURANT

中文翻譯

臺灣臺北市松山路8號　美女餐廳
（02）2202-5948

2015年4月1日

程丹尼
美酒公司
臺灣臺北市信義路502號

親愛的程先生：

感謝您在3月30日來信為內比奧羅紅酒報價。

很遺憾通知您我們公司無法向您下訂單。因為你們商品的價格遠高於其他交易者的報價。若您能降低商品報價，請告訴我。

感謝您的關注。我們期待收到您的消息。

您忠誠的，

范莉莉

范莉莉
總經理
美女餐廳

✓ 內文佳句

1. Dear ① _____ ,
   親愛的 _____ ,

2. Thank you for your quotation dated ②_____ for ③_____ .
   感謝您在 _____ 來信為 _____ 報價。

3. We regret to inform you that our company is unable to place an order.
   很遺憾通知您我們公司無法向您下訂單。

4. Because your prices are much higher than those I have been quoted by other dealers.
   因為你們商品的價格遠高於其他交易者的報價。

5. Please let me know if you can reduce your quoted prices.
   若您能降低商品報價,請告訴我。

6. Thank you for your attention.
   感謝您的關注。

7. We look forward to hearing from you.
   我們期待收到您的消息。

8. ④ Faithfully yours,
   您忠誠的,

▶ 關鍵詞句

1 quotation 報價	**n.** 名詞
2 wine 紅酒	**n.** 名詞
3 regret 遺憾	**v.** 動詞
4 inform＝tell 告知	**v.** 動詞
5 company 公司	**n.** 名詞
6 unable 無法	**adj.** 形容詞
7 order 訂單	**n.** 名詞
8 because 因為	**conj.** 連接詞
9 dealer 經銷商	**n.** 名詞
10 higher 較高的	**adj.** 形容詞（比較級）
11 quote 報價	**v.** 動詞
12 reduce 降低	**v.** 動詞
13 look forward to＋Ving 期待	
14 hear 聽到	**v.** 動詞
15 good news 好消息	**n.** 名詞

▶ 同義詞

1 regret＝be sorry 遺憾

2 be unable to＝be incapable of 無法

3 know＝realize 意識

4 reduce＝diminish 減少

5 look forward to＝hope 期望

書信範例

GOOD WINE
No. 502, Xinyi Rd.,
Taipei, Taiwan
(02) 2122-1388

March 30, 2015

Lily Fan
BEAUTY RESTAURANT
No. 8, Song San Rd.,
Taipei, Taiwan

Dear Ms. Fan,

Thank you for your reply of ① March 25. We are sorry to hear that you consider our prices are too high.

We decide to offer you a special discount of ② 10% on the first order for ③ NTD $ 100,000. We make this allowance because we want to do business with you. We can assure you that the quality of our products and service will meet your expectations.

We hope you will place an order, after reviewing this revised quotation. If you have any further questions, please contact us as soon as possible.

④ Faithfully yours,

Danny Cheng,
General Manager
GOOD WINE

中文翻譯

美酒公司
臺灣臺北市信義路502號
（02）2122-1388

2015年3月30日

范莉莉
美女餐廳
臺灣臺北市松山路8號

親愛的范小姐：

感謝您於3月25日的回覆。我們很遺憾得知您認為我們的價格太高。

我們決定對您金額達100,000元新臺幣的首次訂單，提供10%的特別折扣。我們提供這個優惠，是因為想跟您合作，並能向您保證貨物品質和服務不會因此而打折扣。

我們希望此修正報價能使您下訂單。若有任何問題，請盡速聯絡我們。

您忠誠的，

程丹尼
總經理
美酒公司

✓ 內文佳句

[1] Thank you for your reply of ①_____ .

感謝您於 _____ 的回覆。

[2] We are sorry to hear that you consider our prices are too high.

我們很遺憾得知您認為我們的價格太高。

[3] We decide to offer you a special discount of ②_____ on the first order for ③_____ .

我們決定對您決定對您金額達 _____ 的首次訂單提供 _____ 的特別折扣。

[4] We make this allowance because we want to do business with you.

我們提供這個優惠，是因為想跟您合作。

[5] We cau assure you that the quality of our products and service will meet your expectations.

我們能向您保證貨物品質和服務不會因此而打折扣。

[6] We hope you will place an order, after reviewing this revised quotation.

我們希望此修正報價能使您下訂單。

[7] If you have any fureher questions, please contact us as soon as possible.

若有任何問題，請盡速聯絡我們。

[8] ④ Faithfully yours,

您忠誠的，

▶ 關鍵詞句

1 reply 回覆	**n.** 名詞
2 decide 決定	**v.** 動詞
3 offer 提供	**v.** 動詞
4 special 特別的	**adj.** 形容詞
5 discount 折扣	**n.** 名詞
6 first order 首次訂單	**n.** 名詞
7 allowance 優惠，減價	**n.** 名詞
8 business 生意	**n.** 名詞
9 assure 保證	**v.** 動詞
10 quality 品質	**n.** 名詞
11 service 服務	**n.** 名詞
12 revise 修正	**v.** 動詞
13 enable 使…… 能夠……	**v.** 動詞
14 as soon as possible 盡速	**adv.** 副詞
15 manager 經理	**n.** 名詞

▶ 同義詞

1 think = consider 認為

2 decide = determine 決定

3 allowance = discount 折扣

4 be revised = be modified 修改的

5 assure = guarantee 保證

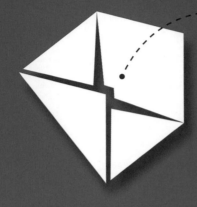

Part 3
簽約成交
與履約

在貿易實務程序中，稱為接受並確認，也稱是簽定契約。買方在接受報價，再經賣方最後確認，買賣契約即告成立。買賣契約一旦成立，雙方的權利義務也因而確定，賣方須交付貨品，買方則須支付貨款。

書信範例

BEAUTY RESTAURANT
No. 8, Song San Rd.,
Taipei, Taiwan
(02) 2202-5948

April 5, 2015

Danny Cheng
GOOD WINE
No. 502 Xinyi Rd.,
Taipei, Taiwan

Dear Mr. Cheng,

Thank you for your quotation dated ①April 2 for ②Nebbiolo wines. Please supply the following items:

Quantity	Product	Item number	Unit price
20	Nebbiolo（1970）	OIM302	NTD＄5,400
10	Nebbiolo（1965）	OIM303	NTD＄6,175
50	Nebbiolo（1960）	OIM304	NTD＄6,300
	Total：＄484,750 NTD		

Payment terms will be standard ③2%-10/NET 30. Please ship the items as soon as possible by ④BCEX expedited shipping. And please deliver all items no later than ⑤April 30, 2015.

Thank you for your handling of our order. If you have any questions, please contact us as soon as possible.

⑥Faithfully yours,

Lily Fan,
General Manager
BEAUTY RESTAURANT

中文翻譯

美女餐廳
臺灣臺北市松山路8號
(02) 2202-5948

2015年4月5日

程丹尼
美酒公司
臺灣臺北市信義路502號

親愛的程先生：

感謝您在4月2日來信為內比奧羅紅酒報價。請供應下列商品：

數量	商品	標號	單價
20	內比奧羅（1970）	OIM302	NTD＄5,400
10	內比奧羅（1965）	OIM303	NTD＄6,175
50	內比奧羅（1960）	OIM304	NTD＄6,300
	總額：新臺幣＄484,750		

若於10天內付款，將享2%的折扣；或是30天內將款項付清。
請盡速以BCEX緊急快遞發送商品。並於2015年4月30日以前寄送所有商品。

謝謝您處理我們的訂單。有任何問題，請盡速聯絡我們。

您忠誠的，

范莉莉
總經理
美女餐廳

商用英語立可貼

✓ 內文佳句

1. Thank you for your quotation dated ①_____ for ②_____ .
   感謝您在 _____ 來信為 _____ 報價。

2. Please supply the following items.
   請供應下列商品：

3. Payment terms will be standard ③X %-Y／NETZ .
   若於 Y_____ 天內付款，將享 X%_____ 的折扣；或是 __ Z_____ 天內將款項付清。

4. Please ship the items as soon as possible by ④_____ .
   請盡速以 _____ 發送商品。

5. Please deliver all items no later than ⑤_____ .
   並於 _____ 以前寄送所有商品。

6. Thank you for your handling of our order.
   謝謝您處理我們的訂單。

7. If you have any question, please contact us as soon as possible.
   有任何問題，請盡速聯絡我們。

8. ⑥ Faithfully yours,
   您忠誠的，

✓ 片語

1. ...... no later than ...... 不遲於……
   Please deliver all items no later than _____ .
   並於 _____ 以前寄送所有商品。

▶ 關鍵詞句

1 購物清單：

數量 Quantity	商品 Product	標號 Item number	單價 Unit price
100	書 book	OLK582	NTD ＄300
30	手機 cell phone	ASM253	NTD ＄9,300
250	電腦 computer	XTR744	NTD ＄36,500
		Total 總價：新台幣 ＄9,434,000	

2 付款方式：

2%-15／NET 45：若於15天內付款將享2%的折扣；若於45天內付款，須付全額。

COD（cash on delivery）貨到付款。

COD not accepted：不接受貨到付款。

3 送貨方式：

Ocean freight service 海運、**Air freight service** 空運、**Road freight service** 陸運

4 賣方寄貨期限（日期：美式）：**June 1**（6月1日）、**May 26**（5月26日）、**April 15**（4月15日）

5 item 商品，項目   **n.** 名詞

6 quantity 數量   **n.** 名詞

7 standard 標準   **n.** 名詞

8 freight 運費、貨物   **n.** 名詞

書信範例

GOOD WINE
No. 502, Xinyi. Rd.,
Taipei, Taiwan
(02) 2122-1388

April 15, 2015

Lily Fan
BEAUTY RESTAURANT
No. 8, Song San Rd.,
Taipei, Taiwan

Dear Ms. Fan,

Thank you for your order of ① April 5. We are pleased to confirm your order as ② the following list.

Quantity	Product	Item number	Unit price
20	Nebbiolo（1970）	OIM302	NTD＄5,400
10	Nebbiolo（1965）	OIM303	NTD＄6,175
50	Nebbiolo（1960）	OIM304	NTD＄6,300
	Total：NTD＄484,750		

We will provide our services at all times. We will inform you as soon as the consignment is ready for ③ transport.

We look forward to hearing from you again. Please let me know if you need any more help.

④ Yours faithfully,

Danny Cheng
General Manager
GOOD WINE

中文翻譯

美酒公司
臺灣臺北市信義路502號
(02) 2122-1388

2015年4月15日

范莉莉
美女餐廳
臺灣臺北市松山路8號

親愛的范小姐：

感謝您於4月5日下訂單。我們很高興向您確認下列的購買商品：

數量	商品	標號	單價
20	內比奧羅（1970）	OIM302	NTD＄5,400
10	內比奧羅（1965）	OIM303	NTD＄6,175
50	內比奧羅（1960）	OIM304	NTD＄6,300
	總額：新臺幣＄484,750		

我們將隨時為您服務。若商品準備好可以出貨，我們將會立刻通知您。

我們期待能再次為您服務。若您需要任何協助，請讓我知道。

您忠誠的，

程丹尼
總經理
美酒公司

✓ 內文佳句

1. Thank you for your order of ①_____ .
   感謝您於 _____ 下訂單。

2. We are pleased to confirm your order as ②_____ .
   我們很高興向您確認 _____ 的購買商品：

3. We will provide our services all the times.
   我們將隨時為您服務。

4. We will inform you as soon as the consignment is ready for ③_____
   _____ .
   若商品準備好可以 _____ ，我們將會立刻通知您。

5. We look forward to hearing from you again.
   我們期待能再次為您服務。

6. Please let me know if you need any more help.
   若您需要任何幫助，請讓我知道。

7. ④Yours faithfully,
   您忠誠的，

✓ 片語

1. ...... be pleased to ...... 很高興……
   We are pleased to confirm your order as _____ .
   我們很高興向您確認 _____ 購買商品。

▶ 關鍵詞句

1. list 表   **n.** 名詞
2. quantity 數量   **n.** 名詞
3. product 商品   **n.** 名詞
4. item number 商品編號   **n.** 名詞
5. unit price 單價   **n.** 名詞
6. total 總額   **n.** 名詞
7. assure 保證   **v.** 動詞
8. receive 獲得   **v.** 動詞
9. notify 通知   **v.** 動詞
10. at once 立刻   **adv.** 副詞
11. consignment 託運物   **n.** 名詞
12. transport 出貨   **n.** 名詞
13. look forward to 期待   **v.** 動詞

▶ 同義詞

1. confirm = acknowlegde 確認
2. quantity = amount 數量
3. notify = inform 通知
4. at once = immediately 立刻
5. transport = shipping 運送

書信範例

GOOD PHONE
No. 502, Xinyi Rd.,
Taipei, Taiwan
(02) 2122-1388

April 15, 2015

Lily Fan
BEAUTY RESTAURANT
No. 8, Song San Rd.,
Taipei, Taiwan

Dear Ms. Fan,

Thank you for your order of ① April 14. However, we regret to tell you that we are unable to fill your order of ② Jack Cell Phone.

Jack has recently ceased production of cell phones. Therefore, we are unable to provide you with ③ Jack Cell Phone. We are very sorry about that.

We recommend you ④ Ace Cell Phone as it is similar to ⑤ the Jack products in terms of function and price.

Please feel free to contact me if you have any question. We will be glad to recommend other products for your specific needs.

⑥ Faithfully yours,

Danny Cheng
General Manager
GOOD PHONE

中文翻譯

好電話公司
臺灣臺北市信義路502號
（02）2122-1388

2015年4月15日

范莉莉
美女餐廳
臺灣臺北市松山路8號

親愛的范小姐：

感謝您在4月14日下訂單。但是很遺憾的，我們無法處理您訂購杰克手機的訂單。

杰克已於近期內停止生產手機，所以我們無法提供您杰克手機。我們對此感到非常抱歉。

我們向您推薦愛斯手機，其功能和價格皆與杰克的產品相似。

若您有任何問題，歡迎與我們聯絡。我們將很樂意為您推薦其他符合您需求的商品。

您忠誠的，

程丹尼
總經理
好電話公司

商用英語立可貼

1. Thank you for your order of ①_____ .
   感謝您在 _____ 下訂單。

2. However, we regret to tell you that we are unable to fill your order of ②_____ .
   但是很遺憾的，我們無法處理您訂購 _____ 的訂單。

3. Therefore, we are unable to provide you with ③_____ . We are very sorry about that.
   所以我們無法提供您 _____。我們對此感到非常抱歉。

4. We recommend you ④_____ as it is similar to ⑤_____ in terms of function and price.
   我們向您推薦 _____，其功能和價格皆與 _____ 相似。

5. Please feel free to contact me if you have any question.
   若您有任何問題，歡迎與我們聯絡。

6. We will be glad to recommend other products for your specific needs.
   我們將很樂意為您推薦其他符合您需求的商品。

7. ④ Faithfully yours,
   您忠誠的，

☑ 片語

1. ...... be unable to ...... 無法……
   We are unable to provide you with _____ .
   我們無法提供您 _____。

082

▶ 關鍵詞句

1 無法供貨的理由：

out of stock 商品缺貨

ceased production 商品停產

2 賣方推薦的商品：

magazine 雜誌、beer 啤酒、power bank 充電器

3 買方欲購的商品：

books 書、drinks 飲品、laptop 筆記型電腦

4 cell phone 手機	n. 名詞
5 recently 近期地	adv. 副詞
6 cease 停止	v. 動詞
7 production 生產	n. 名詞
8 similar 相似的	adj. 形容詞
9 function 功能	n. 名詞
10 free 隨意	adj. 形容詞
11 glad 高興的	adj. 形容詞
12 specific 特定的	adj. 形容詞
13 provide 提供	v. 動詞
14 recommend 推薦、建議	v. 動詞
15 contact 聯絡	v. 動詞

▶ 同義詞

1 however = nevertheless 可是

書信範例

From: Woody Dolan [woodyd@bigcorp.com]
To: Janet Lee [janetl@gigastone.com]
Cc: Grace Wang [gracew@bigcorp.com]
Subject: Sale Confirmations (Order Number #520)

Dear Ms. Lee,

We would like to express how pleased we were to receive your order of ① 13 November 2016.

We confirm supply of ② 120 toy ducks, 4 toy ponies, and 5 teddy bears at the prices stated in your order No. ③ 520 and will allow ④ 4% special discount on your order worth ⑤ $4,500 or above. Our No. 2206 Sales Confirmation in two originals was airmailed to you. Please sign and return one copy of them for our file.

A letter of credit in our favor covering the said ⑥ toys should be issued immediately. We want to point out that stipulations in the relative L/C must be strictly correspondent to those stated in our Sales Confirmation to avoid subsequent amendments. We assure that we will deliver your shipment without any delay on receipt of your letter of credit.

We appreciate your cooperation and look forward to receiving further orders from you.

⑦ Sincerely yours,

Woody Dolan
Sales Manager
Big Corp.

中文翻譯

寄件人：伍迪達蘭 [woodyd@bigcorp.com]
收件人：李珍妮特 [janetl@gigastone.com]
副本：　王格蕾絲 [gracew@bigcorp.com]
主旨：　銷售確認（訂單號碼520）

親愛的李女士，

我想告訴您我們是多麼高興收到您2016年11月13日的訂單。

我們確定以第520號訂單標示的價格，供應您120隻玩具鴨、4隻玩具馬和5隻泰迪熊，同時滿4,500元或以上金額的訂單將獲得4%的特別優惠。我們已把兩份第2206號銷售確認書正本以航空郵件寄給您，請簽署並寄回一份供我們做紀錄。

請立即為我們開立有關支付上述玩具款項的信用狀。我們希望，有關信用狀的規定必須嚴格地與銷售確認書的規定保持一致，以避免日後修改。請放心，我們一收到您的信用狀就會馬上出貨。

謝謝您的配合，希望未來會再收到您的訂單。

您誠摯的，

伍迪達蘭
業務經理
巨大公司

✓ 內文佳句

1. We would like to express how pleased we were to receive your order of ① _____ .

   我想告訴您我們是多麼高興收到您 _____ 的訂單。

2. We confirm supply of ② _____ at the prices stated in your order No. ③ _____ and will allow ④ _____ % special discount on your order worth ⑤ $ _____ or above.

   我們確定以第 _____ 號訂單標示的價格，供應您 _____ __，同時滿 _____ 元或以上金額的訂單將獲得 _____ %的 特別優惠。

3. A letter of credit in our favor covering the said ⑥ _____ should be issued immediately.

   請立即為我們開立有關支付上述 _____ 款項的信用狀。

4. Please sign and return one copy of them for our file.

   請簽署並寄回一份供我們做紀錄。

5. We assure that we will deliver your shipment without any delay on receipt of your letter of credit.

   我們一收到您的信用狀就會馬上出貨。

6. We appreciate your cooperation and look forward to receiving further orders from you.

   謝謝您的配合，希望未來會再收到您的訂單。

7. ⑦ Sincerely yours ,

   您誠摯的，

▶ 關鍵詞句

1 日期（英式）：

13 November（11月3日）、**22 October**（10月22日）

2 貨品：

120 toy ducks, 4 toy ponies, and 5 teddy bears

120隻玩具鴨、4隻玩具馬和5隻泰迪熊

3 allow 允許	**v.** 動詞
4 airmail 航空郵寄	**v.** 動詞
5 sign 簽署	**v.** 動詞
6 return 寄回	**v.** 動詞
7 copy 副本	**n.** 名詞
8 understand 理解，諒解	**v.** 動詞
9 letter of credit＝L/C 信用狀	**n.** 名詞
10 immediately 立即	**adv.** 副詞
11 stipulation 規定	**n.** 名詞
12 relative 相關，有關	**adj.** 形容詞
13 strictly 嚴格地	**adv.** 副詞
14 avoid 避免	**v.** 動詞
15 subsequent 隨後的	**adj.** 形容詞
15 amendment 修改	**n.** 名詞

▶ 同義詞

1 allow＝admit 允許，同意

書信範例

From: Woody Dolan [woodyd@bigcorp.com]
To: Janet Lee [janetl@gigastone.com]
Cc: Grace Wang [gracew@bigcorp.com]
Subject: Contract Terms (S/C No. 1104)

Dear Ms. Lee,

We are glad to know that all ① contract terms have been settled between your company and our sales representative. Please find our ② S/C No. 1104. Please countersign and return for our records.

Our manufacturers informed us that the capacity was fully loaded, so the production must be arranged according to ③ the P/O sequence. Any contract signed after ④ March 5 would not be able to be scheduled for production before ⑤ July 1. Therefore, we urge you to sign and return the contract as soon as you can.

Keep in touch and let me know if you have any question. Thank you.

Sincerely yours,

Woody Dolan
Sales Manager
Big Corp.

中文翻譯

寄件人：伍迪達蘭 [woodyd@bigcorp.com]
收件人：李珍妮特 [janetl@gigastone.com]
副本：　王格蕾絲 [gracew@bigcorp.com]
主旨：　合約條款（第1104號銷售確認書）

親愛的李女士，

我們很高興得知，貴公司跟我方業務代表談妥了所有合約條款。請查收我們第1104號銷售確認書，會簽後請寄回供我們做紀錄。

生產廠商告訴我們產能已滿，因此只能依照接單順序生產。任何在3月5日以後簽署的合約無法在7月1日前排上生產線。因此，我們極力建議你們盡快簽妥合約並寄回給我們。

請保持聯繫，有任何問題請隨時知會，謝謝。

您誠摯的，

伍迪達蘭
業務經理
巨大公司

✓ 內文佳句

⑴ We are glad to know that all ①_____ have been settled between your company and our sales representative.

我們很高興得知，貴公司跟我方業務代表談妥了所有 _____條款。

⑵ Please find our ②_____ . Please countersign and return for our records.

請查收我們 _____，會簽後請寄回供我們做紀錄。

⑶ The capacity was fully loaded, so the production must be arranged according to ③_____ .

產能已滿，因此只能依照 _____ 生產

⑷ Any contract signed after ④_____ would not be able to be scheduled for production before ⑤_____ .

任何在 _____ 以後簽署的合約無法在 _____ 前排上生產線。

⑸ We urge you to sign and return the contract as soon as you can.

我們極力建議你們盡快簽妥合約並寄回給我們。

⑹ Sincerely yours,

您誠摯的，

▶ 關鍵詞句

1 contract terms 合約條款	**n.** 名詞
2 settle 解決	**v.** 動詞
3 company 公司	**n.** 名詞
4 sales representative 業務代表	**n.** 名詞
5 S/C（sales confirmation）銷售確認書	**n.** 名詞
6 countersign 會簽（文件）	**v.** 動詞
7 manufacturer 生產廠商	**n.** 名詞
8 inform 通知	**v.** 動詞
9 urge 極力建議，強烈催促	**v.** 動詞
10 capacity 產能	**n.** 名詞
11 sequence 順序	**n.** 名詞

▶ 同義詞

1 be glad ＝ be happy 高興

2 settle ＝ solve 解決

書信範例

From: Woody Dolan [woodyd@bigcorp.com]
To: Janet Lee [janetl@gigastone.com]
Cc: Grace Wang [gracew@bigcorp.com]
Subject: Modifying Sales Confirmation No. 1515

Dear Ms. Lee,

We have received your letter of ①October 8, 2016 about amending the quantity of the captioned S/C from 1000 doz. to 700 doz. Though ②the quantity was double confirmed by you, we still manage to fulfill your needs. The enclosed Amendment Advice includes ③the quantity amendment and ④expenses incurred. Please sign it back in ⑤two days so we can ⑥rearrange the production in time.

Please be aware that any change to ⑦the S/C would cause interruption to ⑧contract execution. We would appreciate your ⑨re-consideration in the future.

Sincerely yours,

Woody Dolan
Sales Manager
Big Corp.

寄件人：伍迪達蘭 [woodyd@bigcorp.com]
收件人：李珍妮特 [janetl@gigastone.com]
副本：　王格蕾絲 [gracew@bigcorp.com]
主旨：　修改編號1515銷售確認書

親愛的李女士：

我們已收到您2016年10月8日的來信，要求修改主旨訂單數量，從1000打改為700打。雖然當初訂單數量經由您再次確認，但我們依然設法滿足您此次的修改需求。附上修改建議書，其中包含修改數量及因而產生的費用。請於2日內簽回，以便我們及時重新安排生產。

請留意，關於銷售確認書的任何修改，都會造成合約執行上的干擾。將來請您慎重考慮，謝謝。

您誠摯的，

伍迪達蘭
業務經理
巨大公司

要求修改主旨訂單數量，從1000打改為700打

✓ 內文佳句

1. We have received your letter of ①＿＿＿＿＿＿ about amending the quantity of the captioned S/C from 1000 doz. to 700 doz.

   我們已收到您 ＿＿＿＿＿＿ 的來信，要求修改主旨訂單數量，從1000打改為700打。

2. Though ②＿＿＿＿＿＿ was double confirmed by you, we still manage to fulfill your needs.

   雖然當初 ＿＿＿＿＿＿ 經由您再次確認，但我們依然設法滿足您此次的修改需求。

3. The enclosed Amendment Advice includes ③＿＿＿＿＿＿ and ④＿＿＿＿＿＿ .

   附上修改建議書，其中包含 ＿＿＿＿＿＿ 及 ＿＿＿＿＿＿ 。

4. Please sign it back in⑤＿＿＿ so we can ⑥＿＿＿＿＿＿ in time.

   請於 ＿＿＿ 內簽回，以便我們及時 ＿＿＿＿＿＿ 。

5. Any change to ⑦＿＿＿＿＿＿ would cause interruption to ⑧＿＿＿＿＿ .

   關於 ＿＿＿＿＿＿ 的任何修改，都會造成 ＿＿＿＿＿＿ 的干擾。

6. We would appreciate your ⑨＿＿＿＿＿＿ in the future.

   將來請您 ＿＿＿＿＿＿ ，謝謝。

▶ 關鍵詞句

1 amend 修改	**v.** 動詞	
2 S/C 銷售確認書	**n.** 名詞	
3 confirm 確認	**v.** 動詞	
4 fulfill 滿足（願望），履行（義務）	**v.** 動詞	
5 amendment 修改	**n.** 名詞	
6 enclose 附帶	**v.** 動詞	
7 incur（因某事而）產生	**v.** 動詞	
8 cause 導致，造成	**v.** 動詞	
9 interruption 干擾	**n.** 名詞	
10 execution 執行	**n.** 名詞	

▶ 同義詞

1 receive ＝ get 收到

2 amend ＝ change 修改

書信範例

From: Woody Dolan [woodyd@bigcorp.com]
To: Julie Shaw [julies@appleware.com]
Cc: Grace Wang [gracew@bigcorp.com]
Subject: After-sales Service (Order No. 1104)

Dear Ms. Shaw,

①The iPhone 6s, serial No. ②625, which you supplied to us ③last month is out of function. Please refer to our Order No. 1104.

Although our client has followed the instructions in ④operation manual, the phone still did not work well. A ⑤sketch is enclosed for your reference.

Please ⑥take a look into this matter and let us know your technical advice in detail so that ⑦the problem can be solved as soon as possible.

Thank you for your help and cooperation.

Sincerely yours,

Woody Dolan
Sales Manager
Big Corp.

中文翻譯

寄件人：伍迪達蘭 [woodyd@bigcorp.com]
收件人：蕭茉莉 [julies@appleware.com]
副本：　王格蕾絲 [gracew@bigcorp.com]
主旨：　售後服務（1104號訂單）

親愛的蕭女士：

您上月配合我們1104號訂單所供應的iPhone 6s（序號625）發生了故障。

我們的顧客雖然已按照操作說明書中的方法處理，但手機仍然無法正常運作。請參照附上的故障草圖。

請研究這個個案，並給予我們詳盡的技術指導，以便能盡快解決這個問題。

謝謝您的協助與配合。

您誠摯的，

伍迪達蘭
業務經理
巨大公司

商用英語立可貼

1. ① _____ , serial No. ② _____ , which you supplied to us ③ _____ is out of function.

   您 _____ 配合我們 _____ 號訂單所供應的 _____ （序號 _____ ），發生故障。

2. Although our client has followed the instructions in ④ _____ , the phone still did not work well.

   我們的顧客雖然已按照 _____ 中的方法處理，但手機仍然無法正常運作。

3. A ⑤ _____ is enclosed for your reference.

   請參照附上的故障 _____ 。

4. Please ⑥ _____ and let us know your technical advice in detail so that ⑦ _____ as soon as possible.

   請 _____ ，並給予我們詳盡的技術指導，以便能 _____ 。

5. Thank you for your help and cooperation.

   謝謝您的協助與配合。

1. ...... take a look into ...... 調查、研究
   Please take a look into this matter.
   請研究這個個案。

▶ 關鍵詞句

1 serial 序號的	**adj.** 形容詞	
2 client 顧客	**n.** 名詞	
3 follow 跟隨、遵循	**v.** 動詞	
4 instruction 方法，指示	**n.** 名詞	
5 operation 操作	**n.** 名詞	
6 manual 說明書	**n.** 名詞	
7 sketch 草圖	**n.** 名詞	
8 technical 技術的	**adj.** 形容詞	
9 advice 指導	**n.** 名詞	
10 detail 細節	**n.** 名詞	

▶ 同義詞

1 supply ＝ provide 提供

2 remove ＝ eliminate 消除

3 trouble ＝ problem 問題

4 resume ＝ continue 恢復

Part 4
申請開狀
與保險

以信用狀為付款方式,買方必須向往來銀行申請開立信用狀,開狀銀行受理買方之申請後,將依買方指示簽發信用狀。開狀銀行開立信用狀後,一般會透過其在出口地之往來銀行通知信用狀。通知銀行接到國外開狀銀行開來的信用狀,予以查對後,通知信用狀受益人前來領取。另外,在 CIF 貿易條件下,賣方須向保險公司投保貨品運輸保險,取得保險單作為押匯文件。而在 FOB、CFR 貿易條件下,通常會要求買方在申請開狀前先投保貨品運輸保險。

書信範例

From: Winnie Li [winneli@bigcorp.com]
To: Morgan Wang [citibank@citibank.com]
Subject: Open a L/C (Order No. 123)

Dear Mr. Wang,

I am an ①automobile importer. Recently I have ordered limited edition ②cars from ③Taiwan Tech, New York. Please draw a ④Letter of Credit at sight in favor of ⑤Taiwan Tech, New York for ⑥USD$1 million. Please make sure that it is valid to the company until ⑦January 1 and the shipment covering ten cars under order No. 912. Please note that ⑧the insurance gives full cover before accepting the draft.

Thank you for your close cooperation with us in this matter.

Sincerely yours,

Winnie Li
Purchasing Manager
Big Corp.

中文翻譯

寄件人：李維尼 [winneli@bigcorp.com]
收件人：王墨根／花旗銀行 [citibank@citibank.com]
主旨：　申請開發信用狀（訂單編號123）

親愛的王先生，

我是汽車進口商。最近我訂購了一批紐約臺灣科技公司的限量車款。請開立一張金額為100萬美元即期支付的信用狀給紐約臺灣科技公司。確保它是用以支付該公司訂單912號所裝運的10輛汽車，有效期限至1月1日。請注意，在接受該匯票前，保險公司提供全部保障。

感謝您在此事上與我們密切配合。

您最誠摯的，

李維尼
採購經理
巨大公司

商用英語立可貼

1. I am a/an ①＿＿＿＿＿＿ importer.

   我是 ＿＿＿＿＿＿ 進口商。

2. Recently I have ordered limited edition ②＿＿＿＿＿ from ③＿＿＿＿＿＿＿＿ .

   最近我訂購了一批 ＿＿＿＿＿＿ 的限量 ＿＿＿＿＿＿。

3. Draw a ④＿＿＿ at sight in favor of ⑤＿＿＿＿＿＿ for ⑥＿＿＿＿＿ .

   請開立一張金額為 ＿＿＿＿＿ 即其支付的 ＿＿＿＿給 ＿＿＿＿＿。

4. It is valid to the company until ⑦＿＿＿＿＿ ,

   有效期限至 ＿＿＿＿＿＿，

5. Please note that ⑧＿＿＿＿＿＿＿ ,

   請注意，＿＿＿＿＿＿＿＿。

6. Thank you for your close cooperation with us in this matter.

   感謝您在此事上與我們密切配合。

7. Sincerely yours,

   您誠摯的，

✓ 片語

1. ...... in favor of ...... 有利於……

   Please draw a ＿＿＿＿＿＿ at sight in favor of ＿＿＿＿＿＿ for ＿＿＿＿＿ .

   請開立一張金額為 ＿＿＿＿即期支付的 ＿＿＿＿ 給 ＿＿＿＿＿。

▶ 關鍵詞句

1 信用狀分類（依兌現期限）：

Deferred Payment 延期付款、Acceptance Credit 承兌信用

2 issue 開立	**v.** 動詞
3 L/C = letter of credit 信用狀	**n.** 名詞
4 importer 進口商	**n.** 名詞
5 limited 限量	**adj.** 形容詞
6 edition 版本	**n.** 名詞
7 tech 科技	**n.** 名詞
8 sure 肯定	**adj.** 形容詞
9 available 可用的	**adj.** 形容詞
10 until 直至	**p.p.** 介系詞
11 shipment 裝運	**n.** 名詞
12 insurance 保險	**n.** 名詞
13 before 之前	**p.p.** 介系詞
14 accept 接受	**v.** 動詞
15 draft 匯票	**n.** 名詞
16 cooperation 配合	**n.** 名詞

▶ 同義詞

1 car = vehicle 車

2 shipment = delivering 出貨、送貨

3 make sure = ensure 確保

書信範例

From: Grace Yang [graceyang@abcelectronics.com]
To: Paul [paul121@dmv.com]
Subject: Insurance Inguiry

Dear Paul,

We are a supplier of ① smart phones and ② electronic products in ③ Taiwan and the ④ American Chamber of Commerce in ⑤ Taiwan recommended your company to us.

We are interested in receiving the following information:

1. Insurance types
2. Price lists
3. Payment terms

Please send us the information requested above by ⑥ November 25th.

We appreciate your help.

⑦ Regards,

Grace Yang
Sales Manager
ABC Electronics

中文翻譯

寄件人：楊葛雷絲 [graceyang@abcelectronics.com]
收件人：保羅 [paul121@dmv.com]
主旨：　詢問投保費用

親愛的保羅，

我們是智慧型手機和電子產品的臺灣供應商，在臺灣的美國商會向我們推薦貴公司。

我們希望能得到以下的相關資訊：

1. 保險種類
2. 價格表
3. 付款條款

請您在11月25日前提供我們上述提及的資訊。

感謝您的協助。

問候，

楊葛雷絲
業務經理
ABC電子

✓ 內文佳句

1. We are a supplier of ①_____ and ②_____ in ③_____ .

   我們是 _____ 和 _____ 的 _____ 供應商。

2. The ④_____ in ⑤_____ recommended your company to us.

   在 _____ 的 _____ 向我們推薦貴公司。

3. We are interested in receiving the following information:

   我們希望能得到以下的相關資訊：

4. Please send us the information requested above by_____ .

   請您在_____前提供我們上述提及的資訊。

5. ⑥Regards,

   問候，

✓ 結尾佳句：感謝

1. Thank you for your（help／time／assistance／support）.

   謝謝您的（協助／時間／幫助／支持）。

2. Thank you again for your help in this matter.

   再次謝謝您對這件事的幫忙。

▶ 關鍵詞句

1 automotive parts 汽車零件

2 clothing and textile industry 服裝和紡織工業

3 Trade union 工會

4 Industrial Union of Marine and Shipbuilding 海事造船產業

5 supplier 供應商　　　　　　　　　　**n.** 名詞

6 smart phone 智慧型手機　　　　　　　**n.** 名詞

7 electronic 電子的　　　　　　　　　　**adj.** 形容詞

8 chamber 議會　　　　　　　　　　　　**n.** 名詞

9 commerce 商會　　　　　　　　　　　**n.** 名詞

10 recommend 推薦　　　　　　　　　　**v.** 動詞

11 information 資訊　　　　　　　　　　　**n.** 名詞

12 insurance 保險　　　　　　　　　　　**n.** 名詞

13 payment 付款　　　　　　　　　　　　**n.** 名詞

14 term 條件、項目、條款　　　　　　　**n.** 名詞

15 convenient 方便　　　　　　　　　　**adj.** 形容詞

▶ 同義詞

1 supplier = provider 供應者

2 recommend = suggest 推薦

3 request = demand 要求

4 send = transmit 傳送

5 help = assistance 幫忙

書信範例

From: Tony Yang [tonyang@abcfurniture.com]
To: Paul [paul121@dmv.com]
Subject: Favorable Rate of Insurance

Dear Paul,

We are one of the world's top ①furniture exporter and always rent a ②whole cargo liner to ship our goods to ③Taiwan every season. We would like to know whether you are willing to issue the All Risk Insurance for us, and if possible, would you please provide us a favorable insurance rate for ④our regular shipment.

We look forward to your reply .

⑤Best regards,

Tony Yang
Sales Manager
ABC Furniture

中文翻譯

寄件人：楊湯尼 [tonyyang@abcfurniture.com]
收件人：保羅 [paul121@dmv.com]
主旨：　詢問保險特別費率

親愛的保羅，

我們是家具的世界級主要出口商，每季都會租一整班貨輪運貨到臺灣。
想知道貴公司是否樂意為我們的船貨提供全險保單。如果可以，希望能
為我們的定期貨運給予優惠保險費率。

期待回信。

誠摯的問候，

楊湯尼
業務經理
ABC家具

✓ 內文佳句

1 We are one of the the world's top ① _____ exporters.

我們是 _____ 的世界級主要出口商。

2 We always rent a ② _____ to ship our goods to ③ _____ __ every season.

我們每季都會租 _____ 運貨到 _____。

3 We would like to know whether you are willing to issue the All Risk Insurance for us.

我們想知道貴公司是否樂意為我們的船貨提供全險保單。

4 If possible, would you please provide us a favorable insurance rate for ④ _____ .

如果可以,希望能為 _____ 給予優惠保險費利率。

5 We look forward to your reply.

期待回信。

6 ⑤Best regards,

誠摯的問候,

✓ 結尾佳句:未來展望合作

1 I look forward to(hearing from you soon╱meeting you next Tuesday).

期待(您的佳音╱下星期二與您見面)。

2 I look forward to seeing you soon.

期待早日相見。

▶ 關鍵詞句

1	cargo aircraft 貨機	n. 名詞
2	container car 貨櫃車	n. 名詞
3	leading 主要的	adj. 形容詞
4	exporter 出口商	n. 名詞
5	furniture 家具	n. 名詞
6	whole 整個的	adj. 形容詞
7	cargo liner 貨物班輪	n. 名詞
8	season 季	n. 名詞
9	whether 是否	conj. 連接詞
10	willing 樂意的	adj. 形容詞
11	issue 提供、開立	v. 動詞
12	all-risk policy 全損保險	n. 名詞
13	possible 可能的	adj. 形容詞
14	favorable 優惠的	adj. 形容詞
15	regular 定期的	adj. 形容詞

▶ 同義詞

1 be willing ＝ be ready to 樂意

2 issue ＝ grant 授與

3 favorable rate ＝ preferential rate 優惠費率

4 regular ＝ periodical 定期的

5 reply ＝ answer 答覆

書信範例

From: Nancy Liu [nancyliu@sweetfruit.com]
To: Paul [paul121@dmv.com]
Subject: Insurance for the Consignment

Dear Paul,

We have pleasure to inform you that ① we have booked S.S. "Super Star", which is scheduled to leave ② Taiwan for ③ New York on ④ October 23 2015, to carry ⑤ 5,000 cartons of peaches with a total invoice value of ⑥ US $50,000. ⑦ Zack Co., Ltd., the consignee, entrusts us with insuring the consignment against all risks. The amount to be covered is the total invoice value plus ⑧ 10%. We would appreciate if you could quote us the most favorable rate for the coverage.

We look forward to hearing from you soon.

Faithfully yours,

Nancy Liu
Sales Manager
Sweet Fruit Co.

中文翻譯

寄件人：劉南西 [nancyliu@sweetfruit.com]
收件人：保羅 [paul121@dmv.com]
主旨：　貨品投保

親愛的保羅：

很高興通知您，我們已經訂了將於2015年10月23日從臺灣開往紐約的「超級星」號貨輪艙位，運載5,000箱水蜜桃，總額價值五萬美金。收貨人查克有限公司委託我們為這批貨物投保全險。保額為發票總額加10%。若能提供最低保險費率報價，我們會很感激。

期待您的佳音。

您忠誠的，

劉南西
業務經理
甜果公司

✓ 內文佳句

1. We have pleasure to inform you that ①_____ .
   很高興通知您，_____。

2. Is scheduled to leave ②_____ for ③_____ on ④___
   _____ .
   將於 _____ 從 _____ 開往 _____。

3. To carry ⑤_____ with a total invoice value of ⑥_____.
   運載 _____，總額價值_____。

4. ⑦_____ entrusts us with insuring the consignment against all risks.
   _____ 委託我們為這批貨物投保全險。

5. The amount to be covered is the total invoice value plus ⑧____%.
   保額為發票總額加 _____ %。

6. We would appreciate if you could quote us the most favorable rate for the coverage.
   若能提供最低保險費利率報價，我們會很感激。

7. Faithfully yours,
   您忠誠的，

✓ 結尾佳句

1. If I can be of assistance, please do not hesitate to contact me.
   若有我可以協助的，請盡管與我們聯絡。

2. If you require any further information, feel free to contact.
   若您需要任何資訊，隨時聯絡。

▶ 關鍵詞句

1 inform 通知	**v.** 動詞
2 carry 運載，帶	**v.** 動詞
3 carton 箱	**n.** 名詞
4 total 總共	**adj.** 形容詞
5 invoice 單據	**n.** 名詞
6 value 價值	**n.** 名詞
7 consignee 收貨人	**n.** 名詞
8 entrust 委託	**v.** 動詞
9 insure 投保	**v.** 動詞
10 consignment 託運物	**n.** 名詞
11 risk 風險	**n.** 名詞
12 amount 數量	**n.** 名詞
13 plus 加	**p.p.** 介系詞
14 quote 報價	**v.** 動詞

▶ 同義詞

1 inform＝tell 告知

2 carry＝transport 運載

3 consignee＝reciever 收貨人

4 entrust＝authorize 委託

5 quote＝offer 報價

6 be happy to 很高興＝be glad to 樂於＝be willing to 很樂意
＝has／have pleasure to

書信範例

From: Nancy Liu [nancyliu@sweetfruit.com]
To: Paul [paul121@dmv.com]
Subject: Application Form for Insurance

Dear Paul,

Thank you for the letter on ①October 25, 2015 mentioning ②the insurance rate of All Risks.There is no doubt that the rate is very suitable for us. Therefore, we decide to do business with you. Please send us the ③Application form as soon as possible.

We look forward to hearing from you.

④Best regards,

Nancy Liu
Sales Manager
Sweet Fruit Co.

中文翻譯

寄件人： 劉南西 [nancyliu@sweetfruit.com]
收件人： 保羅 [paul121@dmv.com]
主旨： 取得投保申請表

親愛的保羅：

感謝您2015年10月25日的來信提到有關全險的費率。該保險費率毫無疑問非常適合我們。因此，我們決定和貴公司往來。請盡快發送申請表給我。

誠摯的問候，

劉南西
業務經理
甜果公司

商用英語立可貼

✓ 內文佳句

1. Thank you for the letter on ①_____ ,
   感謝您 _____ 的來信，

2. mentioning ②_____ .
   提到有關 _____ 。

3. There is no doubt that the rate is very suitable for us.
   該保險費率毫無疑問非常適合我們。

4. Therefore, we decide to do business with you.
   因此，我們決定和貴公司往來。

5. Please send us the ③_____ as soon as possible.
   請盡快發送 _____ 給我。

6. ④Best regards,
   誠摯的問候，

✓ 結尾佳句

1. Please let me know if you have any questions.
   若您有任何問題，請讓我知道。

2. I hope the above is useful to you.
   期望以上資料對您有幫助。

3. Should you need any further information, please do not hesitate to contact me.
   若您需要任何資訊，請盡管與我聯絡。

▶ 關鍵詞句

1	from 從	p.p.	介系詞
2	letter 信	n.	名詞
3	mention 提及	v.	動詞
4	rate（利）率，（費）率	n.	名詞
5	all risks 全險	n.	名詞
6	doubt 疑問	n.	名詞
7	suitable 適合	adj.	形容詞
8	therefore 因此	conj.	連接詞
9	decide 決定	v.	動詞
10	with 與	p.p.	介系詞
11	send 發送	v.	動詞
12	application 申請	n.	名詞
13	form 表格	n.	名詞
14	as soon as possible 盡快		

▶ 同義詞

1. mention = talk about 講到
2. be suitable for = be advisable to 適當的、可取的
3. do business with = work together with 生意上的往來
4. therefore = consequently 因此
5. decide = determine 決定

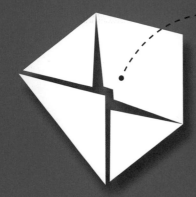

Part 5
包裝與出貨

賣方若發現貨品包裝不一致，需要延遲交貨期；或買方對包裝的反映，缺貨等事宜，都需盡快在製造過程通知對方。賣方在貨品出貨後應發給買方有關貨品詳細資料及裝運情況的通知，讓買方能做好籌措資金、付款、進口通關和提貨的準備。

書信範例

From: Janet Lee [janetl@gigastone.com]
To: Woody Dolan [woodyd@bigcorp.com]
Subject: Package Defects（Order No. 357）

Dear Mr. Dolan,

Our Order No. 357 of 5 laptops arrived this morning. However, one laptop was found badly broken when we ①unwrapped the carton. We noticed that ②one of the outer edges of the wrapping had been worn through, presumably as a result of ③damp in transit.

Please take necessary precautions and make sure that the packing process can entirely protect the goods from ④dampness. These laptops are liable to be ⑤damaged in transit.

Thank you for your cooperation.

⑥Sincerely yours,

Janet Lee
Purchasing Manager
Gigastone.

中文翻譯

寄件人：李珍妮特 [janetl@gigastone.com]
收件人：伍迪達蘭 [woodyd@bigcorp.com]
主旨：　包裝瑕疵（訂單編號357）

親愛的達蘭先生，

訂單編號357的筆記型電腦今天早上已到貨。當我們卸去包裝時發現有一台筆記型電腦遭到嚴重損害。我們注意到包裝的其中一條外邊已被磨穿，大概是運送過程中的潮濕導致。

請採取必要的預防措施，並確保包裝能保護貨品不受潮，因為這些筆記型電腦在運送中容易受到損害。

謝謝您的配合。

您誠摯的，

李珍妮特
採購經理
技佳巨石公司

✓ 內文佳句

1. One laptop was found badly broken when we ①_____.
   當我們 _____ 發現有一台筆記型電腦遭到嚴重損害。

2. We noticed that ②_____ had been worn through.
   我們注意到 _____ 已被磨穿。

3. Presumably as a result of ③_____ in transit.
   大概是運送過程中的 _____ 導致。

4. Please take necessary precautions and make sure that the packing process can entirtly protect the goods from ④_____ .
   請採取必要的預防措施，並確保包裝能保護貨品不 _____ 。

5. These laptops are liable to be ⑤_____in transit .
   因為這些筆記型電腦在運送中容易受到 _____。

6. Thank you for your cooperation.
   謝謝您的配合。

7. ⑧Sincerely yours,
   您誠摯的，

✓ 片語

1. ...... notice that ...... 注意到……
   We noticed that _____ had been worn through.
   我們注意到包裝的其中一條外邊已被磨穿。

▶ 關鍵詞句

1	貨品：five laptops 五台筆記型電腦	
2	可能損害原因：friction or dampness 被摩擦或受潮	
3	contain 裝著	v. 動詞
4	appear 呈現	v. 動詞
5	unwrap 卸去包裝	v. 動詞
6	notice 注意	v. 動詞
7	outer 在外的	adj. 形容詞
8	edge 邊緣	n. 名詞
9	worn 磨損	v. 動詞 Wear 的過去分詞
10	presumably 想必、推測地	adv. 副詞
11	transit 運送	n. 名詞
12	precaution 預防措施	n. 名詞
13	dampness 受潮	n. 名詞
14	liable 容易的	adj. 形容詞
15	spoil 損害	v. 動詞

▶ 同義詞

1	contain ＝ hold 容納
2	appear ＝ show 顯現
3	unwrap ＝ unpack 解開
4	notice ＝ aware 發覺
5	protect ＝ defend 保護

書信範例

From: Woody Dolan [woodyd@bigcorp.com]
To: Janet Lee [janetl@gigastone.com]
Subject: Re: Package Defects（Order No. 357）

Dear Ms. Lee,

We regret to know that ① one of the laptops which you received on ② October 8 was damaged. we are sorry for the inconvenience. We have reported your situation to our insurance company, and made a claim for compensation. We will send you a new ③ laptop as soon as possible.

The new ④ laptop had been examined before being packed and we assure it will be delivered to your company in perfect condition.

Thank you for your understanding.

⑤ Sincerely yours,

Woody Dolan
Sales Manager
Big Corp.

寄件人：伍迪達蘭 [woodyd@bigcorp.com]
收件人：李珍妮特 [janetl@gigastone.com]
主旨：　回覆：包裝瑕疵（訂單編號 357）

親愛的李女士：

我們很遺憾得知您在 10 月 8 日收到的筆記型電腦裡，其中一台損壞了。我們很抱歉造成任何不便。我們已將您的狀況向我們的保險公司報告並申請賠償，而且我們會盡快把新的筆記型電腦寄給您。

新的筆記型電腦在包裝前已經過檢查，我們相信您的貨品一定會完好無缺地送到您公司。

謝謝您的諒解。

您誠摯的，

伍迪達蘭
業務經理
巨大公司

✓ 內文佳句

[1] We regret to know that ①_____ which you received on ②_____ was damaged.

我們很遺憾得知您在 _____ 收到的 _____，其中一台損壞了。

[2] We have reported your situation to our insurance company, and made a claim for compensation.

我們已將您的狀況向我們的保險公司報告並申請賠償。

[3] We will send you a new ③_____ as soon as possible.

我們會盡快把新的 _____ 寄給您。

[4] The new ④_____ had been examined before being packed.

新的 _____ 在包裝前已經過檢查。

[5] We assure it will be delivered to your company in perfect condition.

我們相信您的貨品一定會完好無缺送到您公司。

[6] Thank you for your understaning.

謝謝您的諒解。

[7] ⑤Sincerely yours,

您誠摯的，

✓ 片語

[1] ...... report to ...... 已向……報告

We have reported your claim to our insurance company.

我們已將您提出的索賠要求向我們的保險公司報告了。

▶ 關鍵詞句

1	laptop 筆記型電腦	**n.** 名詞
2	receive 收到	**v.** 動詞
3	deliver 運送	**v.** 動詞
4	damage 損壞	**v.** 動詞
5	report 報告	**v.** 動詞
6	claim 要求	**v.** 動詞
7	insurance 保險	**n.** 名詞
8	send 發送	**v.** 動詞
9	examine 檢查	**v.** 動詞
10	pack 包裝	**v.** 動詞
11	trust 相信	**v.** 動詞
12	reach 到達	**v.** 動詞
13	perfect 完美的	**adj.** 形容詞
14	condition 狀態	**n.** 名詞
15	understanding 諒解	**n.** 名詞

▶ 同義詞

1 dispatch = deliver 發送

2 be damaged = be impaired 被損壞

3 report = inform 報告

4 be examined = be checked 被檢查

5 trust = believe 相信

書信範例

From: Tom [tom001@bigcorp.com]
To: Peter [peterp@gigastone.com]
Subject: Push Order（Order No. 333）

Dear Peter,

We would like to bring your attention to our Order No. ①333 covering ②10 pieces of ③iPhone 6s, for which we sent to you about ④30 days ago an irrevocable L/C expiration date on ⑤October 30th.

When we placed the order, we pointed out that punctual shipment was of the utmost importance because this order was secured from ⑥the largest dealer here and we have given them a definite assurance that we could supply the goods by the end of ⑦October.

We emphasize that any delay in shipping would cause serious problems for us.

We thank you very much for your cooperation.

⑧Best regards,

Tom Williams
Purchasing Manager
Big Corp.

中文翻譯

寄件人： 湯姆 [tom001@bigcorp.com]
收件人： 彼得 [peterp@gigastone.com]
主旨： 催貨（訂單編號 333）

親愛的彼得，

希望您能關注 333 號訂單 10 件 iPhone 6s 的手機，我們在 30 天前發送了無法撤銷的信用狀，而其截止日期為 10 月 30 日。

我們下訂單時已經明確表明，準時出貨是最重要的，因為這份訂單是由這邊最大的經銷商委託的，我們也向他們明確保證能在十月底前供應貨品。

我們想強調任何訂單的出貨延誤，都毫無疑問地會為我們造成嚴重的問題。

我們非常感謝您的配合。

最誠摯的問候，

湯姆威廉斯
採購經理
巨大公司

✓ 內文佳句

1. We would like to bring your attention to our Order No. ①_____ _____ covering ②_____ of ③_____, for which we sent to you about ④_____ days ago an irrevocable L/C expiration date on ⑤_____ .

   希望您能關注 _____ 號訂單 _____ 件 _____的 手機，我們在 _____ 天前發送了無法撤銷的信用狀，而其截 止日期為 _____。

2. When we placed the order, we pointed out that punctual shipment was of the utmost importance.

   我們下訂單時已經明確表明，準時出貨是最重要的。

3. Because this order was secured from ⑥_____ here.

   因為這份訂單是由這邊 _____ 委託的。

4. We have given them a definite assurance that we could supply the goods by the end of ⑦_____ .

   我們也向他們明確保證能在 _____ 月底前供應貨品。

5. We emphasize that any delay in shipping would cause serious problems for us.

   我們想強調任何訂單的出貨延誤，都毫無疑問地會為我們造成嚴重 的問題。

6. We thank you very much for your cooperation.

   我們非常感謝你的配合。

7. ⑧Best regards,

   最誠摯的問候，

▶ 關鍵詞句

1 attention 關注	**n.** 名詞
2 irrevocable 無法撤銷的	**adj.** 形容詞
3 L/C ＝ letter of credit 信用狀	**n.** 名詞
4 expiration date 截止日期	**n.** 名詞
5 punctual 準時的	**adj.** 形容詞
6 utmost 最	**adj.** 形容詞
7 important 重要的	**adj.** 形容詞
8 secure 委託、作保、確保	**v.** 動詞
9 dealer 經銷商	**n.** 名詞
10 definite 明確的	**adj.** 形容詞
11 assurance 保證	**n.** 名詞
12 emphasize 強調	**v.** 動詞
13 undoubtedly 毫無疑問地	**adv.** 副詞
14 involve 造成	**v.** 動詞
15 difficulty 不便	**n.** 名詞

▶ 同義詞

1 assurance ＝ guarantee 保證

2 delay ＝ postpone 延遲

3 emphasize ＝ underline 強調

4 undoubtedly ＝ surely 毫無疑問地

5 difficulty ＝ trouble 困難

書信範例

From: Peter [peterp@gigastone.com]
To: Tom [tom001@bigcorp.com]
Subject: Product Delay

Dear Mr. Williams,

Order No. 333

We are sorry to inform you that it has become impossible to complete shipment before ①the end of October.

In fact, a natural disaster occurred yesterday and a terrible earthquake struck this part of the country on October 13. Our factory suffered from serious damages. As a result, it is impossible to ship before the valid date of the Letter of Credit which expired on October 30th.

Under the circumstances, we hope to extend the credit till ②November 20 as we requested in the ③telephone conversation this morning. ④The delay is beyond our control, the expected shipment will be around ⑤November 10.

At all events, please accept our apologies for the inconvenience you have been put to.

⑥Yours very truly,

Peter Pearson
Manager
Gigastone.

中文翻譯

寄件人：彼得 [peterp@gigastone.com]
收件人：湯姆 [tom001@bigcorp.com]
主旨：　貨品延誤

親愛的威廉斯先生，

訂單號碼333

很抱歉通知您，您要求的訂單不可能如期在10月底出貨。

事實上，昨天發生了一場自然災害，一場可怕的地震在10月13日當天侵襲了部分地方，我們的工廠遭到嚴重破壞，以致無法在信用狀的失效日10月30日前出貨。

在這情況下，如同我們今早在電話裡的請求，希望您會同意延長信用期至11月20日。這次延誤是我們無法控制的。而預計出貨日將大概在11月10日。

無論如何，請接受我們為對您造成的不便而致歉。

您非常真誠的，

彼得皮爾森
經理
技佳巨石公司

商用英語立可貼

✓ 內文佳句

[1] We are sorry to inform you that it has become impossible to complete shipment before①_____ .

很抱歉通知您，您要求的訂單不可能如期在 _____ 出貨。

[2] Under the circumstances, we hope to extend the credit till ②_____ as we requested in the ③_____ .

在這情況下，如同我們 _____ 裡的請求，希望您會同意延長信用期至 _____ 。

[3] ④_____ is beyond our control.

_____ 是我們無法控制的。

[4] The expected shipment will be around ⑤_____ .

而預計出貨日將大概在 _____ 。

[5] At all events, please accept our apologies for the inconvenience you have been put to.

無論如何，請接受我們為對您造成的不便而致歉。

[6] ⑥Yours very truly,

您非常真誠的，

✓ 片語

[1] ...... beyond our control ...... 我們無法控制的……

Although _____ is beyond our control, the expected shipment will be around _____ .

_____ 是我們無法控制的。而預計出貨日將大概在 _____ 。

▶ 關鍵詞句

1	impossible 不可能的	adj. 形容詞
2	complete 完成	v. 動詞
3	natural disaster 自然災害	n. 名詞
4	occur 發生	v. 動詞
5	terrible 可怕的	adj. 形容詞
6	earthquake 地震	n. 名詞
7	struck 侵襲	v. 動詞（過去式）
8	country 國家	n. 名詞
9	factory 工廠	n. 名詞
10	suffer 遭到	v. 動詞
11	serious 嚴重的	adj. 形容詞
12	validity 合法性	n. 名詞
13	circumstance 情況	n. 名詞
14	apology 致歉	n. 名詞
15	inconvenience 不便	n. 名詞

▶ 同義詞

1 complete = fulfil 完成

2 occur = happen 發生

3 terrible = awful 可怕的

4 agree = consent 同意

5 extend = prolong 延（期）

書信範例

From: Grace Lin [gracelin@bigcorp.com]
To: Betty [gbetty@gigastone.com]
Subject: Shipments postponement

Dear Betty,

We confirm the receipt of ①your email dated ②June 4, 2016, requesting to postpone the shipment.

According to the contract No. ③4520 about ④30,000 boxes of drinks, you have not been granted the letter of credit; therefore, the goods cannot be shipped by the originally booked vessel. You asked us to postpone the shipment and rebook ⑤ "Super Power". We agree and have had the L/C amended.

As regards the said contract, it is certain that we issued L/C No. ⑥0236 on ⑦June 21 according to the new schedule. We also have inquired the reason and responsibility of the bank for such a delay in issuing L/C.

We apologize for the inconvenience again.

⑧Sincerely yours,

Grace Lin
Sales Manager
Big Corp.

中文翻譯

寄件人：林格蕾絲 [gracelin@bigcorp.com]
收件人：貝蒂 [gbetty@gigastone.com]
主旨：　通知延期出貨

親愛的貝蒂，

我們確認收到在 2016 年 6 月 4 日您寄出的電子郵件，要求延期發貨。

關於 4520 號合約訂購 30,000 箱飲料一事，由於您尚未取得信用狀，貨品不能依原訂的船隻運送。您要求我們延期發貨，改訂「超級力量」號船運貨。我們同意並已修改信用狀內容。

至於上述合約，根據新的船期，我們確定在 6 月 21 日開出 0236 號信用狀，我們已向銀行追查此次信用狀延誤的原因和責任。

我們再次為造成不便致歉。

您誠摯的，

林格蕾絲
業務經理
巨大公司

✓ 內文佳句

[1] We confirm the receipt of ①_____ dated ②_____ .
我們確認收到在 _____ 您 _____ 。

[2] According to the contract No. ③_____ about ④_____
_____ , you have not been granted the letter of credit; therefore,
the goods cannot be shipped by the originally booked vessel. You
asked us to postpone the shipment and rebook ⑤_____ .
關於 _____ 號合約訂購 _____ 一事，由於您尚未取
得信用狀，貨品不能用原訂船隻運送，您要求我們延期發貨，改訂
_____ 號船運貨。

[3] We agree and have had the L/C amended.
我們同意並已修改信用狀內容。

[4] As regards the said contract, it is certain that we issued L/C No. ⑥
_____ on ⑦_____ according to the new schedule.
至於上述合約，根據新的船期，我們確定在 _____ 開出 ____
_____ 號信用狀。

[5] We also have inquired the reason and responsibility of the bank for
such a delay in issuing L/C.
我們已向銀行追查此次信用狀延誤的原因和責任。

[6] ⑧Sincerely yours,
您誠摯的，

▶ 關鍵詞句

1 acknowledge 確認	**v.** 動詞	
2 receipt 收到	**n.** 名詞	
3 cable 電報	**n.** 名詞	
4 reference 參考	**n.** 名詞	
5 contract 合約	**n.** 名詞	
6 originally 原來地	**adv.** 副詞	
7 vessel 船	**n.** 名詞	
8 postpone 延期	**v.** 動詞	
9 amend 修改	**v.** 動詞	
10 certain 確定的	**adj.** 形容詞	
11 schedule 時間表	**n.** 名詞	
12 receive 收	**v.** 動詞	
13 inquire 追查	**v.** 動詞	
14 responsibility 責任	**n.** 名詞	
15 inform 通知	**v.** 動詞	

▶ 同義詞

1 ask = demand 請求

2 postpone = put off 延期

4 receive = in receipt of 收到

5 cancel = call off 取消

書信範例

From: Woody [woody@bigcorp.com]
To: Amy [amyll@gigastone.com]
Subject: Shipping by Airmail Economy

Dear Amy,

Thank you for your letter of ① October 15 informing us the reason of the shipping delay.

Considering the delivery will be too late to arrive, we request to change delivery method by ② air mail. Airmail Economy is a cheap and safe way of sending goods in packets by ③ air freight so that the consignment will be on time.

Do not worry about the higher shipping fee; we will bear all the extra cost.

I appreciate your early warning.

④ Thanks,

Woody Dolan
Sales Manager
Big Corp.

中文翻譯

寄件人：伍迪 [woody@bigcorp.com]
收件人：艾米 [amyll@gigastone.com]
主旨：　通知空運出貨

親愛的艾米，

感謝您10月15日來信通知我們發貨延誤的原因。

考慮到商品會太晚送抵，我們要求您能改變運送方式，以航空郵件運送，經濟空運是便宜及安全的空運包裝寄貨，以便貨品準時抵達。

不用擔心超支的運費，我們將承擔所有額外費用。

感謝您及早示警。

謝謝，

伍迪達蘭
業務經理
巨大公司

商用英語立可貼

✓ 內文佳句

1. Thank you for your letter of ①_____ informing us the reason of the shipping delay.

   感謝您 _____ 來信通知我們發貨延誤的原因。

2. Considering the delivery will be too late to arrive.

   考慮到商品會太晚送抵。

3. We request to change delivery method by ②_____ .

   我們要求您能改變運送方式，以 _____ 運送。

4. Airmail Economy is a cheap and safe way of sending goods in packets by ③_____ so that the consignment will be on time.

   經濟空運是便宜及安全的 _____ 包裝寄貨，以便貨品準時抵達。

5. Do not worry about the higher shipping fee; we will bear all the extra cost.

   不用擔心超支的運費，我們將承擔所有額外費用。

6. ④Thanks,

   謝謝

✓ 片語

1. ...... too late to ...... 會太晚……

   Considering the delivery will be too late to arrive, we request to change delivery method by _____.

   考慮到商品會太晚送抵，我們要求您能改變運送方式，以 _____ 運送。

▶ 關鍵詞句

1	consider 考慮	v. 動詞
2	consignment 委託貨物	n. 名詞
3	late 晚的	adj. 形容詞
4	arrive 送抵，到達	v. 動詞
5	change 改變	v. 動詞
6	shipping 運送	n. 名詞
7	method 方式	n. 名詞
8	instead of 替代	p.p. 介系詞
9	air mail 空運	n. 名詞
10	worry 擔心	v. 動詞
11	expect 預計	v. 動詞
12	amount 數目	n. 名詞
13	bear 承擔	v. 動詞
14	extra 額外的	adj. 形容詞
15	cost 費用	n. 名詞

▶ 同義詞

1	consider ＝ regard 考慮
2	arrive ＝ reach 到達
3	transit ＝ convey 運送
4	send ＝ deliver 遞送
5	bear ＝ assume 承擔

書信範例

From: Jerry [jerry@bbdrink.com]
To: Peter [peterp@gigastone.com]
Subject: Shipment Notification

Dear Peter,

We are writing to inform you that we have delivered ①three boxes of ②milk tea which you ordered ③yesterday.

The ④sweetness and ⑤storage temperature were adjusted, and we trust ⑥drinks will reach you safely.

Please pay special attention that in order to maintain the quality of the ⑦dairy products, store the milk tea ⑧in a fridge at 5°C. Please ⑨drink them as soon as possible. Never drink the milk tea when it's left at room temperature overnight.

Thank you very much for your cooperation and we look forward to receiving your order again.

⑩Best wishes,

Jerry Lo
Sales Assistant
BB Drink

中文翻譯

寄件人：傑瑞 [jerry@bbdrink.com]
收件人：彼得 [peterp@gigastone.com]
主旨：　通知出貨

親愛的彼得，

我們來信是為了通知您，我們已發送您昨天訂購的三箱奶茶。

您要的甜度和冷藏溫度已調整，我們相信飲料會安全送到您手上。

請特別注意，為了維持乳製品的品質，請將奶茶冷藏於攝氏5度。請盡速喝完，室溫下存放隔夜請勿飲用。

我們非常感謝您的配合，期待您下次的訂單。

最佳的祝福，

羅傑瑞
業務助理
BB飲品

✓ 內文佳句

1. We are writing to inform you that we have delivered ①_____
   of ②_____ which you ordered ③_____ .

   我們來信是為了通知您，我們已發送您 _____ 訂購的 _____
   _____ 。

2. The ④_____ and ⑤_____ were adjusted.

   您要的 _____ 和 _____ 已調整。

3. We trust ⑥_____ will reach you safely.

   我們相信____會安全送到您手上。

4. Please pay special attention that in order to maintain the quality of
   the ⑦_____, store the milk tea ⑧_____.

   請特別注意，為了維持_____的品質，請將奶茶_____。

5. Please ⑨_____ as soon as possible. Never drink the milk tea
   when it's left at room temperature overnight.

   請盡速_____，室溫下存放隔夜請勿飲用。

6. Thank you very much for your cooperation and we look forward to
   receiving your order again.

   我們非常感謝你的配合，期待您下次的訂單。

7. ⑩Best wishes,

   最佳的祝福，

▶ 關鍵詞句

1 Send 寄、transmit 發送		
2 Delivery Service 送貨服務	**n.** 名詞	
3 freight 貨物運輸	**n.** 名詞	
4 dispatch 派遣	**v.** 動詞	
5 road 道路	**n.** 名詞	
6 carrier 送貨員	**n.** 名詞	
7 box 箱	**n.** 名詞	
8 milk tea 奶茶	**n.** 名詞	
9 sweetness 甜度	**n.** 名詞	
10 adjust 調整	**v.** 動詞	
11 reach 到達	**v.** 動詞	
12 safely 安全地	**adv.** 副詞	
13 special 特別的	**adj.** 形容詞	
14 dairy product 乳製品	**n.** 名詞	
15 hours 數小時	**n.** 名詞	
16 finish 完成	**v.** 動詞	

▶ 同義詞

1 adjust ＝ trim 調整

2 safely ＝ in security 安全地

3 maintain ＝ keep 保持

4 store ＝ preserve 保存、儲藏

Part 6

運輸

發貨通知應按契約或信用狀規定的時間發出。賣方可詢問貨運公司運費，到貨時通知買方到達進口地，若不幸發生事故，也須在第一時間通知對方。買方也可發信詢問貨櫃情況等。

書信範例

From: Winnie Li [winniel@gigastone.com]
To: Yi-Jing [yijing@bigmarine.com]
Subject: Shipping Freight Inquiry

Dear Yi-Jing,

In early ① November, we will have ② ten limited edition cars to deliver from ③ Kaohsiung, Taiwan to ④ Dubai. The ⑤ cars will be packed in ⑥ ten Twenty-foot Equivalent Units, each measuring ⑦ 6.1*2.44*2.62m and weighing ⑧ 25.5 tonnes.

Please state:

1. Fees for delivery.
2. Sailing schedule for ⑨ November.
3. Estimated time of sailing.

Thank you for your instant reply.

Sincerely yours,

Winnie Li
Sales Assistant
GIGASTONE

中文翻譯

寄件人：李維尼 [winniel@gigastone.com]
收件人：怡晶 [yijing@bigmarine.com]
主旨：　詢問運費

親愛的怡晶，

11月初，我們將有十輛限量版汽車要從高雄運送到杜拜。汽車將被裝在10個20英吋的標準貨櫃，每個大小為6.1 X 2.44 X 2.62公尺，重25.5公噸。

請說明：

1. 運費
2. 11月的船期表
3. 預計的航期

感謝您的盡速回覆。

您誠摯的，

李維尼
業務助理
技佳巨石公司

商用英語立可貼

1 In early ①_____ ,

_____ 初，

2 we will have ②_____ to deliver from ③_____ to
④_____ .

我們將有 _____ 要從 _____ 運送
到 _____ 。

3 The ⑤_____ will be packed in ⑥_____, each
measuring ⑦_____m and weighing ⑧_____ tonnes.

_____ 將被裝在 _____ 個 _____ 英尺
的標準貨櫃，每個大小為 _____ × _____ × _____
_____ 公尺，重 _____ 公噸。

4 Fees for delivery.

運費。

5 Sailing schedule for⑨ _____.

_____ 的船期表。

6 Estimated time of sailing.

預計的航期。

7 Thank you for your instant reply.

感謝您的盡速回覆。

8 Sincerely yours,

您誠摯的，

156

▶ 關鍵詞句

1 貨櫃：

forty-foot equivalent unit（FEU） 40英呎標準貨櫃

twenty-foot 20英呎

2	beginning 初	n. 名詞
3	limited 限量的	adj. 形容詞
4	edition 版本	n. 名詞
5	pack 包裝	v. 動詞
6	equivalent 等量的	adj. 形容詞
7	each 每	pron. 代名詞
8	tonne 公噸	n. 名詞
9	state 說明	v. 動詞
10	arrange 安排	v. 動詞
11	estimate 預計	v. 動詞
12	customer 客戶	n. 名詞
13	assistance 協助	n. 名詞

▶ 同義詞

1 fee ＝ cost 費用

2 customer ＝ client 顧客

3 estimate ＝ evaluate 估計

書信範例

From: Winnie Li [winniel@gigastone.com]
To: Sara Ma [sarama@firstforward.com]
Subject: Container Service Information

Dear Sara,

We found your company information from the website and knew that you are committed to ①shipping our cargo in safe and efficient manner.

We would be glad if you can send us more detailed information of ②container liner services covering from the ③Far East to ④Europe, and main line trades to ⑤Canada and ⑥America , including the ⑦charge, ⑧equipment or any ⑨prohibited lists.

We are looking forward to hearing from you soon.

Sincerely yours,

Winnie Li
Sales Assistant
GIGASTONE

中文翻譯

寄件人：李維尼 [winniel@gigastone.com]
收件人：馬莎拉 [sarama@firstforward.com]
主旨：　詢問貨櫃服務

親愛的莎拉，

我們從網路上得知貴公司資訊，提供快速貨運運輸安全及快速的服務。

如果您能寄給我們關於服務地域包含遠東到歐洲貨櫃航線以及加拿大和美國航線的詳情，包括收費、貨櫃設備及禁止事項，我們會很感謝您。

期待佳音。

您最誠摯的，

李維尼
業務助理
技佳巨石公司

✓ 內文佳句

1. Dear _____ ,
   親愛的 _____ ,

2. We found your company information from the website and knew that you are committed to ①_____ .
   我們從網路上得知貴公司資訊，提供 _____ 及 _____ 的服務。

3. Information of ②_____ covering from the ③_____ to ④ _____ , and main line trades to ⑤_____ and ⑥_____ ____.
   關於_____包含 _____ 到 _____ 貨櫃航線 以及 _____ 和 _____ 航線的詳情。

4. We would be glad if you can send us more detailed information, including the ⑦_____ , ⑧_____ or any _____ lists.
   如果您能寄給我們相關的詳情，包括____、_____及____事項，我會很感謝您。

5. We are looking forward to hearing from you soon.
   期待佳音。

6. Sincerely yours,
   您最誠摯的，

✓ 片語

1. ......find ......from 從……得知
   We found your company information from the website.
   我們從網路上得知貴公司資訊。

▶ 關鍵詞句

❶ subject 主旨	**n.** 名詞	
❷ container 貨櫃	**n.** 名詞	
❸ service 服務	**n.** 名詞	
❹ understand 知道	**v.** 動詞	
❺ operate 經營	**v.** 動詞	
❻ Netherlands 荷蘭	**n.** 名詞	
❼ route 航線	**n.** 名詞	
❽ should 應該	**aux.** 助動詞	
❾ glad 高興的	**adj.** 形容詞	
❿ send 寄	**v.** 動詞	
⓫ detail 詳情	**n.** 名詞	
⓬ about 有關	**p.p.** 介系詞	
⓭ include 包括	**v.** 動詞	
⓮ charge 收費	**n.** 名詞	
⓯ importer 進口商	**n.** 名詞	

▶ 同義詞

❶ main＝primary 主要的

❷ including＝consist of 包括

❸ charge＝fee 收費

❹ equipment＝device 設備

❺ prohibited＝forbidden 被禁止的

書信範例

From: Sara Ma [sarama@firstforwarder.com]
To: Winnie Li [winniel@gigastone.com]
Subject: Re: Container Service Information

Dear Ms. Li,

Thank you for your inquiry of ① October 10. As for your questions, the shipping containers we provide have ② two sizes:

Product	Max Load	Price
Twenty-foot Equivalent Units	21.6 tonne	$ 2450 (USD)
Forty-foot Equivalent Units	26.5 tonne	$ 4600 (USD)

The containers have great advantages, including both ③ watertight and ④ airtight. They can also be loaded and locked at the ⑤ factory to ensure your cars remaining safe.

We have enclosed our ⑥ price list for your reference. We hope to hear from you soon.

⑦ Yours sincerely,

Sara Ma
Sales Assistant
First Forwarder

中文翻譯

寄件人：馬莎拉 [sarama@firstforwarder.com]
收件人：李維尼 [winniel@gigastone.com]
主旨： 回覆詢問貨櫃服務狀況

親愛的李小姐，

感謝您在10月10日來信詢問，我們提供的海運貨櫃有兩種尺寸：

產品	最大負載	價格
20呎標準貨櫃	21.6噸	2450美元
40呎標準貨櫃	26.5噸	4600美元

貨櫃具有防水和不透氣的優點。而且可以在工廠裡裝運和上鎖，以確保您汽車的安全。

我們附上價目表，希望很快能收到您的回覆。

您誠摯的，

馬莎拉
業務助理
第一貨運

商用英語立可貼

✓ 內文佳句

1. Thank you for your inquiry of ①_____ .
   感謝您在 _____ 來信詢問。

2. The shipping containers we provide have ②_____ sizes:
   我們提供的海運貨櫃有 _____ 種尺寸

3. The containers have great advantages, including both ③_____
   and ④_____ .
   貨櫃具有 _____ 和 _____ 的優點。

4. They can also be loaded and locked at the ⑤_____ to
   ensure your cars remaining safe.
   可以在 _____ 裡裝運和上鎖，以確保您汽車的安全。

5. We have enclosed our ⑥_____ for your reference. We hope
   to hear from you soon.
   我們附上 _____ ，希望很快能收到您的回覆。

6. ⑦Yours sincerely,
   您誠摯的，

▶ 關鍵詞句

1 防護類型：

anti-collision 防撞

high-temperature-resistant 耐熱

deformation resistance 防變形

2	service 服務	n. 名詞
3	inquiry 詢問	n. 名詞
4	ship 運送	v. 動詞
5	provide 提供	v. 動詞
6	max 最大的	adj. 形容詞
7	load 負載	n. 名詞
8	equivalent 等量物	n. 名詞
9	tonne 公噸	n. 名詞
10	advantage 優點	n. 名詞
11	watertight 防水的	adj. 形容詞
12	airtight 不透氣的	adj. 形容詞
13	lock 上鎖	v. 動詞
14	factory 工廠	n. 名詞
15	ensure 確保	v. 動詞
16	safe 安全的	adj. 形容詞

▶ 同義詞

1 load = fill 裝滿

2 advantage = merit 優點

3 ensure = assure 保證

4 factory = plant 工廠

5 be enclosed = be attached 被附上

書信範例

From: Winnie Li [winniel@gigastone.com]
To: John Peterson [jpeterson@triumph.com]
Subject: Goods Arrival Notification

Dear John,

Order No. ①912

We have duly received your products which you dispatched on ②November 11. We unpacked the package and inspected the goods. The contents were in perfect condition. We are very pleased with your service.

As for the rest ③payment for the goods, we have issued the ④L/C at sight, and sent it by courier.

We are looking forward to doing business with you again.

⑤Best regards,

Winnie Li
Sales Assistant
GIGASTONE

中文翻譯

寄件人： 李維尼 [winniel@gigastone.com]
收件人： 約翰彼德森 [jpeterson@triumph.com]
主旨： 通知貨品收到

親愛的約翰，

訂單編號912

我們已如期收到您於11月11日發送的產品。我們打開包裹並檢查裡面的貨品，所有的物品都完好無缺。我們非常滿意您的服務！

至於其餘的貨款，我們已經發出了即期信用狀，並已透過快遞寄出。

我們期待與您進一步交易。

最好的問候，

李維尼
業務助理
技佳巨石公司

商用英語立可貼

① Oder No. ①_____

訂單編號 _____

② We have duly received your products which you dispatched on
②_____ .

我們已如期收到您於 _____ 發送的產品。

③ We unpacked the package and inspected the goods. The contents
were in perfect condition.

我們打開包裹並檢查裡面的貨品，所有的物品都完好無缺。

④ We are very pleased with your service.

我們非常滿意您的服務！

⑤ As for the rest ③_____ for the goods, we have issued
the ④_____ and sent it by courier.

至於其餘的 _____ ，我們已經發出了 _____ ，並已
透過快遞寄出。

⑥ We are looking forward to doing business with you again.

我們期待與您進一步交易。

⑦ ⑤Best regards,

最好的問候，

① ...... be very pleased with ...... 我們非常滿意⋯／我們非常高興⋯⋯
We are very pleased with your service.

我們非常滿意您的服務！

▶ 關鍵詞句

1 支付方式：

Banker's Transfer 銀行轉帳

Bill of Exchange (B/E) 匯票

in advance 預付

on open account 賒帳

2 寄送方式：

by registered airmail 航空掛號信

by courier 快遞

3 notification 通知　　　　　　　　　　**n.** 名詞

4 duly 適時地　　　　　　　　　　　　**adv.** 副詞

5 dispatch 發送　　　　　　　　　　　**v.** 動詞

6 unpack 打開包裹　　　　　　　　　　**v.** 動詞

7 inspect 檢查　　　　　　　　　　　　**v.** 動詞

8 content 內容　　　　　　　　　　　　**n.** 名詞

9 We have duly received _____

我們已如期收到 _____

10 We are very pleased with your work! ＝表示謝意

＝ We appreciate the effort you're making to complete this transaction.

＝ Thank you for your close cooperation with us in this matter.

書信範例

From: Winnie Li [winniel@gigastone.com]
To: Sherry Sung [sherrysung@dylon.com]
Subject: Notification of Transportation Delay（Order No. 567）

Dear Sir,

We regret to inform you that your order No. 567 will be delayed for ①a week. Due to the ②heavy snowstorm, all of the ③railway transports are shut down. Therefore, your products are trapped on the way to the destination.

According to the forecast of Weather Bureau, ④the snowstorm will ease on ⑤this weekend. We will resume your delivery immediately as soon as the weather improves.

Thank you for your patience. Once again, we are sorry for any inconvenience.

⑥Yours faithfully,

Winnie Li
Sales Assistant
GIGASTONE

中文翻譯

寄件人： 李維尼 [winniel@gigastone.com]
收件人： 宋雪莉 [sherrysung@dylon.com]
主旨： 運輸途中事故通知（訂單編號567）

親愛的客戶，

我們很遺憾通知您，您編號567的訂單將會被延誤一個禮拜。由於強烈暴風雪影響，所有的鐵路運輸都被迫中止。因此，您的貨物被困在運送途中。

根據氣象局的預測，暴風雪將在本週末減緩。只要天氣好轉，我們將立即恢復您貨品的運送。

感謝您的耐心。我們再次為帶給您不便表示歉意。

您忠誠的，

李維尼
業務助理
技佳巨石公司

✓ 內文佳句

1 We regret to inform you that your order No. 567 will be delayed for
①_____.

我們很遺憾通知您，您編號567的訂單將會被延誤 _____。

2 Due to the ②_____ , all of the ③_____ transports
are shut down.

由於 _____ 影響，所有的 _____ 運輸都被迫中止。

3 Therefore, your products are trapped on the way to the destination.

因此，您的貨物被困在運送途中。

4 According to the forecast of Weather Bureau, ④_____ will
ease on ⑤_____ .

根據氣象局的預測，_____ 將在 _____ 減緩。

5 We will resume your delivery immediately as soon as the weather
improves.

只要天氣好轉，我們將立即恢復您貨品的運送。

6 Thank you for your patience.

感謝您的耐心。

7 Once again, we are sorry for any inconvenience.

我們再次為帶給您不便表示歉意。

8 ⑥Yours faithfully,

您忠誠的，

▶ 關鍵詞句

1 意外狀況：car crash 車禍、traffic jam 塞車

2 運輸：air transport 空運、sea transport 海運

3 天氣狀況：flood 水災、drought 旱災、thick fog 濃霧

4 局處：Custom 海關、Highway Bureau 公路總局、Weather Bureau 氣象局

5 accident 事故	**n.** 名詞	
6 snowstorm 暴風雪	**n.** 名詞	
7 shut 關閉	**v.** 動詞	
8 trap 困住	**v.** 動詞	
9 destination 目的地	**n.** 名詞	
10 ease 減緩	**v.** 動詞	
11 improve 改善	**v.** 動詞	

12 We regret to inform you that _____

我們很遺憾通知您 _____

13 because of _____ ＝ due to _____

由於 _____

14 We are sorry for any inconvenience.

We are awfully sorry for the inconvenience.

We apologize for any inconvenience this may cause.

＝我們為帶給您不便表示歉意。

在貨品裝船出口後，賣方要遵從信用狀之規定，備齊信用狀規定之匯票、發票、提單，保險單等單據，向往來銀行申請押匯。押匯銀行依指示，將單據寄往開狀銀行或其指定銀行請求付款。買方付款後，開狀銀行即將貨運單據交予買方。

書信範例

From: Winnie Li [winniel@gigastone.com]
To: Ryan Yang [ryanyang@citibank.com]
Subject: Payment by Transfer (L/C No. 912)

Dear Mr. Yang,

I'm Winnie Li, the ①automobile importer. I wrote to you on ②November 11 instructing to issue the L/C No. ③912 for us. Today, I have received the statement from my seller and as requested, I need to implement the outstanding balance of ④USD $ 0.5 million. Please remit the balance for the credit to my seller's account with the ⑤Citibank, New York. I will be in your bank next Monday.

Thank you for your close cooperation with us.

⑥Yours very truly,

Winnie Li
Purchasing Assistant
GIGASTONE

中文翻譯

寄件人： 李維尼 [winniel@gigastone.com
收件人： 楊來恩 [ryanyang@citibank.com]
主旨：　以轉帳付款（編號912的信用狀）

親愛的楊先生，

我是李維尼，汽車進口商。我在11月11日寫信給您，請您開立編號912 的信用狀。今天，我收到賣方寄來的帳單，按照要求，我需要繳付50萬 美元的剩餘款項。請將餘款匯至我賣方於紐約花旗銀行的帳戶。下周一 我會到您銀行辦理。

感謝您密切配合我們。

您非常真誠的，

李維尼
採購助理
技佳巨石公司

✓ 內文佳句

1. I'm Winnie Li, the ①_____ .
   我是李維尼，_____ 進口商。

2. I wrote to you on ②_____ instructing to issue the L/C No.
   ③_____ for us.
   我在 _____ 寫信給您，請您開立編號 _____ 的信用狀。

3. Today, I have received the statement from my seller.
   今天，我收到賣方寄來的帳單。

4. As requested, I need to implement the outstanding balance of
   ④_____ .
   按照要求，我需要繳付 _____ 的剩餘款項。

5. Please remit the balance for the credit to my seller's account with
   the ⑤_____ , _____ .
   請將餘款匯至我賣方於 _____ 的帳戶。

6. Thank you for your close cooperation with us.
   感謝您密切配合我們。

7. ⑥Yours very truly,
   您非常真誠的，

✓ 片語

1. ...... from my seller ...... 從我的賣方……
   Today, I have received the statement from my seller
   今天，我收到賣方寄來的帳單。

▶ 關鍵詞句

1 未清償債務：

existing debt 現有債項

unliquidated claim 未清還索賠

2 buyer 買方	**n.** 名詞
3 instruct 通知	**v.** 動詞
4 remit 匯寄	**v.** 動詞
5 transfer 轉帳	**v.** 動詞
6 credit 信用	**n.** 名詞
7 statement 帳單	**n.** 名詞
8 seller 賣家	**n.** 名詞
9 agree 符合	**v.** 動詞
10 request 要求	**v.** 動詞
11 implement 執行	**n.** 名詞
12 outstanding 未支付的	**adj.** 形容詞
13 balance 差額	**n.** 名詞
14 account 帳戶	**n.** 名詞
15 close 密切的	**adj.** 形容詞
16 matter 事項	**n.** 名詞

▶ 同義詞

1 outstanding balance ＝ unpaid balance 未付差價

2 be requested ＝ be asked 被請求

書信範例

From: Winnie Li [winniel@gigastone.com]
To: Woody Dolan [woodyd@bigcorp.com]
Subject: Payment by L/C（Order No. 370）

Dear Mr. Dolan,

According to the specification and price list you provided, the ① cars we ordered are ready to ship. We have informed ② Citibank, New York, to issue the L/C for ③ USD $1 million in your favor, effective until ④ January 31. The L/C will be confirmed by ⑤ Citibank, Taipei, who will accept your draft at ⑥ 90 days for the full amount of your invoice. They will require the following shipping documents to be attached to your draft:

Bill of Lading in duplicate,
Invoice, CIF New York in triplicate,
Insurance for USD $ 0.9 million

Once the details of shipment are confirmed, please inform us.

We appreciate the effort you made for the service.

⑦ Best Regards,

Winnie Li
Purchasing Assistant
GIGASTONE

中文翻譯

寄件人：李維尼 [winniel@gigastone.com]
收件人：伍迪達蘭 [woodyd@bigcorp.com]
主旨：　通知以信用狀付款（訂單編號370）

親愛的達蘭先生，

根據您提供的規格和價格表，我們訂購的車已備妥、可出貨。我們已通知了紐約花旗銀行，開立貴公司為抬頭的100萬美元信用狀，到期日為1月31日。該信用狀會由臺北花旗銀行進行確認，該銀行將會在90天內接受你們的全額匯票。他們將要求您於匯票裡附加以下的運輸文件：

提單一式兩份
發票一式三份，包含成本、保險、到紐約的運費，
價值90萬美元的保險單

一旦確定裝船的細節，請通知我們。

我們很感謝您為完成此事所付出的努力。

最好的問候，

李維尼
採購助理
技佳巨石公司

✓ 內文佳句

[1] According to the specification and price list you provided, the
①＿＿＿＿＿＿ we ordered are ready to ship.

根據您提供的規格和價格表，我們訂購的 ＿＿＿＿＿＿ 已備妥、可出貨。

[2] We have informed ②＿＿＿＿＿＿＿＿ , to iissue the L/C for
③＿＿＿＿＿＿ in your favor, effective until ④＿＿＿＿＿＿ .

我們已通知了 ＿＿＿＿＿＿ ，開立貴公司為抬頭的 ＿＿＿＿＿＿ 信
用狀，到期日為 ＿＿＿＿＿＿ 。

[3] The L/C will be confirmed by ⑤＿＿＿＿＿＿ , who will accept your
draft at ⑥＿＿＿＿＿ days for the full amount of your invoice.

該信用狀會由 ＿＿＿＿＿＿ 進行確認，該銀行將會在 ＿＿＿＿＿＿
天內接受你們的全額匯票。

[4] They will require the following shipping documents to be attached
to your draft:

他們將要求您於匯票裡附加以下的運輸文件：

[5] ⑦Best Regards,

最好的問候，

✓ 片語

[1] ...... effective until ...... 有效期至……

We have informed ＿＿＿＿＿＿ , to issue the L/C for ＿＿＿＿＿＿
in your favor, effective until ＿＿＿＿＿＿ .

我們已通知了 ＿＿＿＿＿＿ ，開立貴公司為抬頭的 ＿＿＿＿＿＿ 信
用狀，到期日為 ＿＿＿＿＿＿ 。

▶ 關鍵詞句

1 裝船單據：

commercial invoice 商業發票

certificate of origin 產地證明

packing list 裝箱表

2 成交價格：

FOB destination 目的地交貨

FOB shipping point 起運點交貨

C&F 成本＋運費

3 specification 規格   **n.** 名詞

4 effective 有效的   **adj.** 形容詞

5 draft 匯票   **n.** 名詞

6 require 要求   **v.** 動詞

7 document 文件   **n.** 名詞

8 We have received your _____

我們已經收到了您 _____

9 The credit will be confirmed by _____

該信用狀會由 _____ 進行確認

10 Once _____ ,

一旦 _____

▶ 同義詞

1 specification ＝ standard 標準、規格

書信範例

From: Daniel [Daniel@happydress.com]
To: Andy Chen [Andy@gigastone.com]
Cc: Cary Lin [Cary@happydress.com]
Subject: Reference Order No. 00054

Dear Mr. Chen,

Reference Order No. 00054

We are pleased to inform you that the ① dresses of your Order ② No. 00054 on ③ December 12 are ready to deliver. Our policy with new customers is full payment ④ 10 days before all shipment, preferably via ⑤ bank transfer. We also accept ⑥ credit card payment.

Thank you for your valuable support to ⑦ Happy Dress. If you have any question, please let me know.

⑧ Best regards,

Daniel Chou
Sales Assistant
Happy Dress

中文翻譯

寄件人：丹尼爾 [Daniel@happydress.com]
收件人：陳安迪 [Andy@gigastone.com]
副本：　林卡里 [Cary@happydress.com]
主旨：　關於訂單編號00054

親愛的陳先生：

關於訂單編號00054

很高興通知您，您12月12日訂單編號00054的洋裝已可以出貨。根據本公司政策，新客戶需先在裝運前十日付清全額款項，並且最好能透過銀行轉帳付款。我們也接受信用卡付款。

感謝您對快樂洋裝寶貴的支持。如有任何問題，請讓我知道。

最誠摯的問候，

周丹尼爾
業務助理
快樂洋裝

✓ 內文佳句

1. We are pleased to inform you that the ①_____ of your Order ②_____ on ③_____ are ready to deliver.

   很高興通知您，您 _____ 訂單編號 _____的_____
   已可以出貨。

2. Our policy with new customers is full payment ④_____ before all shipment.

   根據本公司政策，新客戶需先在裝運前 _____ 付清全額款
   項。

3. preferably via ⑤_____ .

   最好能透過 _____付款 。

4. We also accept ⑥_____ .

   我們也接受 _____。

5. Thank you for your valuable support to ⑦_____ .

   感謝您對 _____ 寶貴的支持。

6. If you have any question, please let me know .

   如有任何問題，請讓我知道。

7. ⑧Best regards,

   最誠摯的問候，

✓ 片語

1. Thank you for ...... 感謝……
   Thank you for your valuable support to Happy Dress.
   感謝您對快樂洋裝寶貴的支持。

▶ 關鍵詞句

1 訂單號碼：

No. 00011 ＝ Number 00011 ＝編號00011 ＝ 00011號

2 訂單的日期（美式）：

December 12（12月12日）

May 1st, 2016（2016年5月1日）

July 26th of last year（去年7月26日）

3 付清款項的期限：

within 10 days 10天內，10 days before shipment 出貨前10天

within 30 days 30天內

within one week 1週內

4 付款方式：

bank transfer 銀行轉帳付款

credit card payment 信用卡付款

wire transfer 電匯付款

cash 現金付款

a letter of credit 信用狀

5 寫信人公司（或品牌）名稱：

Happy Dress 快樂洋裝

Mood Music 心情音樂

London Mood Music 倫敦心情音樂公司

6 結尾語：

Faithfully yours 您忠實的

Sincerely yours 您誠摯的

Best regards 最誠摯的問候

7	pleased 高興的	adj. 形容詞
8	supply 供應	v. 動詞
9	dress 洋裝	n. 名詞
10	include 包含	v. 動詞
11	policy 政策	n. 名詞
12	customer 客戶	n. 名詞
13	require 要求	v. 動詞
14	preferably 最好地	adv. 副詞
15	via 透過	p.p. 介系詞
16	transfer 轉帳	v. 動詞
17	accept 接受	v. 動詞
18	valuable 寶貴的	adj. 形容詞
19	patronage 支持	n. 名詞
20	additional 其他的	adj. 形容詞

▶ 關鍵詞句

1 in advance＝ahead of time 提早

2 customer＝shopper 顧客

3 shipment＝loading 貨載

4 purchase＝transaction 交易

5 bank transfer＝remittance 匯款

書信範例

From: Daniel [Daniel@happydress.com]
To: Andy Chen [Andy@gigastone.com]
Cc: Cary Lin [Cary@happydress.com]
Subject: Reference Order No. 00037

Dear Mr. Chen,

Reference Order No. 00037

According to our records, we have not yet received the ①L/C of your Order No. ②00037, placed on ③December 17, 2015. Please expedite the L/C to us as soon as you can, so that the shipment can be delivered without delay. In order to avoid ④subsequent amendment, please make sure that the ⑤L/C stipulations should be in exact accordance with the terms in ⑥Sales Confirmation.

We hope to hear from you soon.

⑦Best regards,

Daniel Chou
Sales Assistant
Happy Dress

中文翻譯

寄件人： 丹尼爾 [Daniel@happydress.com]
收件人： 陳安迪 [Andy@gigastone.com]
副本： 林卡里 [Cary@happydress.com]
主旨： 關於訂單編號00037

親愛的陳先生：

關於訂單編號00037

根據本公司的紀錄，我們還未收到貴公司於2015年12月17日訂單編號00037之信用狀。請盡快開立，以便我們能順利執行裝運。為避免以後修改，請務必確定信用狀的規定完全按照銷售確認書的條款。

希望很快能收到您的消息。

最誠摯的問候，

周丹尼爾
業務助理
快樂洋裝

商用英語立可貼

✓ 內文佳句

[1] According to our records, we have not yet received the ① _____ of your Order No. ②_____ , placed on ③_____ .

根據本公司的紀錄，我們還未收到貴公司於 _____
訂單編號 _____ 之_____。

[3] Please expedite the L/C to us as soon as you can, so that the shipment can be delivered without delay.

請盡快開立，以便我們能順利執行裝運。

[4] In order to avoid ④_____ , please make sure that the ⑤_____ stipulations should be in exact accordance with the terms in ⑥_____ .

為避免 _____，請務必確定 _____ 的規定完全按照 _____ 的條款。

[5] We hope to hear from you soon.

希望很快能收到您的消息。

[6] ⑦Best regards,

最誠摯的問候，

✓ 片語

[1] According to our records ...... 根據我們的紀錄……
According to our records, we have not yet received the L/C of your Order No. 0037, placed on December 17, 2015.

根據本公司的紀錄，我們還未收到貴公司於 2015 年 12 月 17 日訂單編號 00037 之信用狀。

Part 8
申訴與索賠

商務活動中，針對對方索賠函中提出的索賠意願和要求而回覆的申訴處理信函。從樹立良好形象及未來的業務拓展角度而言，無論企業還是個人，在收到索賠函後都應當認真對待、及時處理，以誠懇友好的態度處理問題，有錯認錯，無則嘉勉。通常來說，理賠函包括以下主要條款：1、說明來函目的；2、糾紛處理態度及意見；3、賠償處理意見；4、感謝並期待未來繼續合作。

書信範例

From: Andy Chen [andy@gigastone.com]
To: customer service [customer-service@happydress.com]
Cc: Grace Lin [gracel@gigastone.com]
Subject: Claim For P/O No. 00037

Dear ① Customer Service Representative,

It is with regret that I must inform you of some serious problems with your delivery of ② four boxes of dresses, ③ No. 00037, which was received on ④ May 1st, 2016.

After carefully examining, the dresses we received are unequal in quality to the sample pieces. The quality is so poor that we feel that there must be some mistakes in the order.

Since the contract is concluded under the term of "⑤ sales by sample", any discrepancy between the sample and the goods under the contract is unacceptable. We have no choice but to ask you to take them back and replace them with goods as good as the sample. If this is not possible, I am afraid we shall have to cancel our order and ask for refund.

⑥ Best regards,

Andy Chen
Sales Manager
Gigastone

中文翻譯

寄件人：陳安迪 [andy@gigastone.com]
收件人：客戶服務 [customer-service@happydress.com]
副本：　林格蕾絲 [gracel@gigastone.com]
主旨：　訂單編號00037 索賠

親愛的客戶服務代表，

很遺憾必須通知您，貴公司於2016年5月1日送達的四箱洋裝（訂單號碼00037）出現嚴重的問題。

經仔細檢查，你們運來給我們的洋裝，品質與訂購時的樣品不符。有些品質太差，讓我們覺得你們在準備這批訂貨時，一定是有些地方弄錯了。

由於合約條款裡表明「憑樣品交易」，樣品跟貨品的任何差異都是不能接受的。我們實在沒有其他辦法，只好要求你們把貨退回，並換成符合訂購品質的貨物。如果不可能的話，恐怕我們得撤銷訂單並要求退款。

最誠摯的問候，

陳安迪
業務經理
技佳巨石公司

✓ 內文佳句

1. Dear ①_____ ,
   親愛的 _____

2. It is with regret that I must inform you of some serious problems with your delivery of ②_____ ,③_____ , which was received on ④_____ .
   很遺憾必須通知您，貴公司於 _____ 送達的 _____
   （訂單號碼 _____）出現嚴重的問題。

3. After carefully examining, the dresses we received are unequal in quality to the sample pieces. The quality is so poor that we feel that there must be some mistakes in the order.
   經仔細檢查，你們運來給我們的洋裝，品質與訂購時的樣品不符。有些品質太差，讓我們覺得你們在準備這批訂貨時，一定是有些地方弄錯了。

4. Since the contract is concluded under the term of "⑤_____" , any discrepancy between the sample and the goods under the contract is unacceptable.
   由於合約條款裡表明「 _____ 」，樣品跟貨品的任何差異都是不能接受的。

5. We have no choice but to ask you to take them back and replace them with goods as good as the sample. If this is not possible, I am afraid we shall have to cancel our order and ask for refund.
   我們實在沒有其他辦法，只好要求你們把貨退回，並換成符合訂購品質的貨物。如果不可能的話，恐怕我們得撤銷訂單並要求退款。

6. ⑥Best regards,
   最誠摯的問候，

▶ 關鍵詞句

1 訂購的貨物名稱及數量：

four boxes of dresses 四箱洋裝

ten LCD screens 十台液晶螢幕

one Bench Lathe 一個桌上型車床

2 貨物到貨日期（美式）：

May 1st, 2016（2016年5月1日）

July 26th of last year（去年7月26日）

3 合同內的條件：

sales by sample 憑樣品交易

書信範例

From: customer service [customer-service@happydress.com]
To: Andy Chen [Andy@gigastone.com]
Cc: Grace Lin [Grace@gigastone.com]
Subject: Claim for P/O No. 00037

Dear ①Mr. Chen,

Thank you for your email explaining the problem with your order ②No. 00037, which is on ③May 1st, 2016.

However, ④the goods have been in your possession for past ⑤3 months and were just unwrapped on ⑥August 1st of this year. We have to refer you to ⑦clause 9 therein, which provides that "in case of quality discrepancy, the Buyer shall file his claim with Seller ⑧within thirty days after arrival of the goods at the port of destination". We regret to inform you that your claim cannot be accepted as it is far beyond the time limit in the contract.

We understand how frustrated it is to have things happened like this. We are unable to offer you any refund, yet we would be happy to offer you a ⑨10% discount off on your next order.

As always, we value our business relationship with you. If I can be of any other assistance, please do not hesitate to contact me.

⑩Best regards,

Daniel Wu
Customer Service Representative
Happy Dress

中文翻譯

寄件人：客戶服務 [customer-service@happydress.com]
收件人：陳安迪 [Andy@gigastone.com]
副本：　林格蕾絲 [grace@gigastone.com]
主旨：　訂單編號00037索賠

親愛的陳先生，

謝謝您來信說明於2016年5月1日送達之訂單編號00037商品問題。

然而，由於商品過去3個月來一直在您的手上，直到今年8月1日才開箱。我們必須請您參考合約第9條，規定「萬一貨物品質不符，買方應在貨物送抵目的地後30天內向賣方提出索賠」。很抱歉通知您，我們不會受理您的索賠，因為這已經遠遠超過合約規定的索賠期限。

不過，我們也了解這樣的事有多令人沮喪。雖然無法為您辦理退款，但是我們很樂意在您下一次訂購時提供您10%的折扣。

我們一如既往地重視這份合作關係。如果還有任何我可以提供協助的地方，請讓我知道。

最誠摯的問候，

吳丹尼爾
客戶服務代表
快樂洋裝

商用英語立可貼

1 Dear ①_____ ,
親愛的 _____

2 Thank you for your email explaining the problem with your order ②_____ , which is on ③_____ .
謝謝您來信說明於 _____ 送達之訂單編號 _____ 商品問題。

3 However, ④_____ have been in your possession for past ⑤_____ and were just unwrapped on ⑥_____ .
然而，由於 _____ 過去 _____ 來一直在您的手上，直到 _____ 才開箱。

4 We have to refer you to ⑦_____ therein, which provides that "in case of quality discrepancy, the Buyer shall file his claim with Seller ⑧_____ after arrival of the goods at the port of destination".
我們必須請您參考合約 _____，規定「萬一貨物品質不符，買方應在貨物送抵目的地後 _____ 內向賣方提出索賠」。

5 We regret to inform you that your claim cannot be accepted as it is far beyond the time limit in the contract.
很抱歉通知您，我們不會受理您的索賠，因為這已經遠遠超過合約規定的索賠期限。

6 We understand how frustrated it is to have things happened like this.
不過，我們也了解這樣的事有多令人沮喪。

7 We are unable to offer you any refund, yet we would be happy to offer you a ⑨_____ on your next order.

雖然無法為您辦理退款，但是我們很樂意在您下一次訂購時提供您 _____。

8 As always, we value our business relationship with you. If I can be of any other assistance, please do not hesitate to contact me.

我們一如既往地重視這份合作關係。如果還有任何我可以提供協助的地方，請讓我知道。

9 ⑩Best regards,

最誠摯的問候，

▶ 關鍵詞句

1 訂單之商品替換字：

goods 貨物，boxes 箱子，clothes 衣物

2 商品於顧客手上的時間：

3 weeks 3週，3 months 3個月，2 years 2年

3 合約條約款項：

clause 9 第9條，section 10 第10節

4 到貨後可提出索賠之期限：

within 30 days 30天內

within 10 days 10日內

within one week 1週內

5 優惠折扣：

10% discount off 10%的折扣，5% discount 5%的折扣

書信範例

From: customer service [customer-service@bigcorp.com]
To: Andy Chen [Andy@gigastone.com]
Cc: Cary Lin [Cary@bigcorp.com]
Subject: Reference No. 0504

Dear ①Mr. Chen,

Thank you for your letter of ②December 21st complaining about the ③USB Drives, Order ④No. 0504, which has caused us a great deal of concern. Thank you again for bringing this to our attention.

We have come up with a solution to test the ⑤USB Drives from the same production batch and agreed that ⑥they cannot be accessed. The defect has been traced to ⑦a fault in one of the machines.

Please return the faulty goods to us, and we will send you ⑧ the new USB Drives right away.

We would like to express our deepest apologies for the inconvenience.

⑨Best regards,

Sunny Wen
Customer Service Representative
Big Corp.

中文翻譯

寄件人： 客戶服務 [customer-service@bigcorp.com]
收件人： 陳安迪 [Andy@gigastone.com]
副本： 林卡里 [Cary@bigcorp.com]
主旨： 訂單編號 0504

親愛的陳先生，

感謝您在 12 月 21 日來信投訴訂單編號 0504 所供應的隨身碟，引起了我們極大的關注。謝謝您讓我們注意到這些問題。

我們對同一生產批號的隨身碟做了檢測，同意他們無法讀取數據。已查明這些缺陷是其中一台機器發生問題引起的。

請把有問題的商品退回給我們，我們馬上寄給您新的隨身碟。

我們為對您造成的麻煩表示最深的歉意。

最誠摯的問候，

溫桑妮
客戶服務代表
巨大公司

✓ 內文佳句

1. Dear ①_____ ,
   親愛的 _____

2. Thank you for your letter of ②_____ complaining about the ③_____ , Order ④_____ , which has caused us a great deal of concern.
   感謝您在 _____ 來信投訴訂單編號 _____ 所供應的 _____ ，引起了我們極大的關注。

3. Thank you again for bringing this to our attention.
   謝謝您讓我們注意到這些問題。

4. We have come up with a solution to test the ⑤_____ from the same production batch and agreed that ⑥_____ . The defect has been traced to ⑦_____ .
   我們對同一生產批號的 _____ 做了檢測，同意他們 _____ 。已查明這些缺陷是 _____ 引起的。

5. Please return the faulty goods to us, and we will send you ⑧_____ right away.
   請把有問題的商品退回給我們，我們將會馬上寄給您 _____ 。

6. We would like to express our deepest apologies for the inconvenience.
   我們為對您造成的麻煩表示最深的歉意。

7. ⑨Best regards,
   最誠摯的問候，

▶ 關鍵詞句

1 訂單的日期（美式）：

December 21st（12月21日）

May 1st, 2016（2016年5月1日）

July 29th of last year（去年7月29日）

2 商品缺陷問題：

they cannot be accessed 它們無法讀取數據

they are broken 它們壞了

their contents are spoiled or stained

包裝內貨物被汙損或弄髒

3 商品缺陷的原因：

a fault in one of the machines

其中一台機器出現問題

4 商品數量：

one hundred USB drives 100個USB隨身碟

ten cameras 10台相機

one TV screen 一台電視螢幕

5 結尾語：

Best regards, 最誠摯的問候

Faithfully yours, 您忠誠的

Sincerely yours, 您誠摯的

書信範例

From: Woody Dolan [woodyd@bigcorp.com]
To: Mia Su [mia@toplaw.com]
Cc: Grace Wang [gracew@bigcorp.com]
Subject: Reference No. 00037

Dear ① Miss Su,

We have difficulties with ② D. Curry, New York, concerning a consignment of ③ wood shipped to them in ④ December. They argued ⑤ the wood was defected and refused to pay. We should be grateful if you would act as an arbitrator in this matter. You can find details in the enclosed statements.

Our main concern is that ⑥ the extremely minor defect should be acceptable. We have already consulted a reputable and well-known manufacturer, who supported our arguement. We thought our customer's claim is unreasonable, and perhaps raised in order to obtain a reduction in the contract price.

We hope you will agree to act for us and come to an understanding with our customer's arbitrator.

⑦ Best regards,

Woody Dolan
Sales Manager
Big Corp.

中文翻譯

寄件人： 伍迪達蘭 [woodyd@bigcorp.com]
收件人： 蘇蜜婭 [mia@toplaw.com]
副本： 王格蕾絲 [gracew@bigcorp.com]
主旨： 訂單編號00037

親愛的蘇小姐，

我們與紐約的D・科里公司有糾紛，事關在12月運送給他們的一批木材。他們質疑我們供應的木材有瑕疵，並拒絕付款。如果您能為此事擔任仲裁，我們將非常感激。細節可以從附件聲明中了解。

我們主要的論點是極小的瑕疵應該是可接受的。我們已請教了一位有信譽且知名的製造業者，他支持我們的論點。因此，我們認為客戶的異議是不合理的，也許提出的目的是為了降低合約價格。

希望您接受我們的委任並向客戶的仲裁取得諒解。

最誠摯的問候，

伍迪達蘭
業務經理
巨大公司

✓ 內文佳句

[1] Dear ①_____ ,
親愛的 _____

[2] We have difficulties with ②_____ , concerning a consignment of ③_____ shipped to them in ④_____ .
我們與 _____ 公司有糾紛，事關在 _____ 運送給他們的一批 _____ 。

[3] They argued ⑤_____ and refused to pay.
他們質疑我們供應的 _____ ，並拒絕付款。

[4] We should be grateful if you would act as an arbitrator in this matter. You can find details in the enclosed statements.
如果您能為此事擔任仲裁，我們將非常感激。細節可以從附件聲明中了解。

[5] Our main concern is that ⑥_____ should be acceptable.We have already consulted a reputable and well-known manufacturer, who supported our arguement.
我們主要的論點是 _____ 應該是可接受的。我們已請教了一名有信譽且知名的製造業者，他支持我們的論點。

[6] We thought our customer's claim is unreasonable, and perhaps raised in order to obtain a reduction in the contract price.
因此，我們認為客戶的異議是不合理的，也許提出的目的是為了降低合約價格。

[7] We hope you will agree to act for us and come to an understanding with our customer's arbitrator.
希望您接受我們的委任並向客戶的仲裁取得諒解。

⑧ ⑦Best regards,

最誠摯的問候，

▶ 關鍵詞句

1 買方的公司名稱：

D. Curry, New York 紐約的D・科里公司

R. Martin, Nairobi 奈洛比的R・馬丁公司

2 買賣的商品：

wood 木材、cotton textiles 棉織品、steels 鋼材

3 買方所提出的質疑：

the wood is defected 木材的瑕疵

the quality of the textiles 棉織品的品質

the corrosion of steel 鋼材鏽蝕

4 商品遇到的問題：

defect 缺陷、corrosion 生鏽、damage 損壞

5 雙方約定的期限：

three days 3天、a week 1週、seven days 7天

附錄

快速成交
商用英文
書信範例

本章附上 19 篇書信往來範例，讓各位讀者可以快速找到所需的情境、直接套用，便可快速寫出一封英文書信。

CCC
No. 43, Sec. 4,
Keelung Rd.,
Da'an Dist., 106
Taipei City, Taiwan R.O.C
Tel: 02-2237-1566 Fax: 02-2237-7135

Dear friends,

We are pleased to inform you that CCC is now opened at No. 43, Sec. 4, Keelung Rd, Da'an District, Taipei City, Taiwan.

We pride ourselves on our complete and diverse lines of 3C products. We specialize in products designed for all ages. We also supply computer accessories and various kinds of phone cases. Besides, we are pleased to offer 10% discount to new customers.

Our employees are trained and eager to find suitable 3C products for you. Enclosure, for your review, is a partial list of our current stocks. We invite feedback from customers, if there is something you would like us to supply, please inform us and we will do our best to fulfill your need.

What are you waiting for? Call us now!

Best regards,

Yi-Ching Chen
Manger
CCC

Enclosure 1

中文翻譯

臺灣臺北市（106）大安區基隆路4段43號　CCC公司
電話：02-2237-1566　傳真：02-2237-7135

親愛的朋友：

我們很高興告訴您CCC現於臺灣臺北市大安區基隆路4段43號正式開張。

我們對本公司完整且多元化的電子產品感到自豪。我們專注於為所有年齡的客人設計產品。我們還供應電腦配件及各種手機外殼。此外，我們很樂意為新顧客提供10%的折扣。

我們的員工受過訓練並樂意為您尋找合適的電子產品。附件裡是我們部份現有商品庫存清單，可供您查閱。我們邀請顧客提出反饋，如果您希望本公司提供某些產品，請通知我們，我們會盡力滿足您的需求。

還等什麼？現在就打電話給我們吧！

陳依菁
經理
CCC公司

附件1

書信範例

CCC
No. 43, Sec. 4,
Keelung Rd.,
Da'an Dist., 106
Taipei City, Taiwan, R.O.C
Tel: (886)02-2237-1566 Fax:(886)02-2237-7135

October 2, 2015

APPLE
No. 1, Infinite Loop,
Cupertino, CA 95014

Dear Sir,

I am the purchasing manager of CCC, Taiwan.
I would like to receive price information for iPhone 6s 16 GB and iPhone 6s 32GB. In addition, what kinds of accessories will be included if I purchase?

Please feel free to contact me at 02-2337-1566 or via email at ccc@gmail.com. I look forward to hearing from you.

Best wishes,

Yi-Ching Chen
Purchasing Manager
CCC

中文翻譯

臺灣臺北市（106）大安區基隆路4段43號　CCC公司
電話：02-2237-1566　傳真：02-2237-7135

2015年10月2日

加州（95014）庫比提諾無窮環路1號　蘋果公司

親愛的先生：

我是CCC在臺灣的交易經理，我想收到iPhone 6s 16GB手機和iPhone 6s 32GB手機的價格資訊。此外，如果我願意購買，會附上什麼配件呢？

如有任何問題，請隨時撥02-2337-1566或透過電子郵件ccc@gmail.com聯繫我。期待您的佳音。

陳依菁
採購經理
CCC

書信範例

APPLE
No. 1, Infinite Loop,
Cupertino, CA 95014

October 10, 2015

CCC
No. 43, Sec. 4,
Keelung Rd.,
Da'an Dist., 106,
Taipei City, Taiwan R.O.C

Ref: Q172301

Dear Ms. Chen,

Thank you for your inquiry of 2nd October. I am pleased to provide you with the enclosed product and the price list for our iPhone 6s 16GB and iPhone 6s 32GB. We are happy to offer you a 5% discount on payments paid within 15 days of delivery.

We require cash in advance for the first time order. We accept bank transfers, check or credit card payments over the phone. We will deliver your products outside US by EMS.

To place an order, please complete the form attached and send me back via email.

Please don't hesitate to contact me if you need any further information.

Yours sincerely,

Jacky Snow
Sales Manager
APPLE

Enclosures 2

中文翻譯

加州（95014）庫比提諾無窮環路1號　蘋果公司

2015年10月10日

臺灣臺北市（106）大安區基隆路4段43號　CCC公司

參考編號：Q172301

親愛的陳小姐，

感謝您在10月2日的詢價。我很高興為您附上我們iPhone 6s 16GB手機和iPhone 6s 32GB手機的產品及價目表。本公司很樂意為您提供在送貨15天內付款的5%折扣優惠。

我們要求在第一次訂貨時先付款。我們接受轉帳、支付支票或透過電話以信用卡付費。我們使用國際快捷運貨到美國以外的地區。

如需下訂單，請填妥附上的表單並以電郵回傳。

如果您需要其他資訊，請馬上聯絡我。

傑克・斯諾
銷售經理
蘋果公司

附件2

下訂單

（正式商業書信寫作文體）

CCC
No. 43, Sec. 4,
Keelung Rd.,
Da'an Dist., 106,
Taipei City, Taiwan, R.O.C
Tel: (886) 02-2237-1566 Fax: (886) 02-2237-7135

October 16th, 2015

APPLE
No. 1, Infinite Loop,
Cupertino, CA 95014

Ref: Order Number #333

Attention: Mr. Jacky Snow

Please supply the following items:

Quantity	items	Item Number	Unit Price
5	iPhone 6s (16GB) Golden	QA002	NTD$25,900
5	iPhone 6s (64GB) Golden	QA001	NTD$29,500
			Total: NTD$277.000

Payment terms shall be standard 5%-10/NET 30.
Please ship the products as soon as possible using EMS.
Please deliver all products before October 30th, 2015.

Ship all items to:

CCC
No. 43, Sec. 4,
Keelung Rd.,
Da'an Dist., 106,
Taipei City, Taiwan, R.O.C.
Tel: (886) 02-2237-1566 Fax: (886) 02-2237-7135

Regards,

Yi Ching Chen
Purchasing Manager
CCC

中文翻譯

臺灣（R.O.C）臺北市（106）大安區基隆路4段43號　CCC公司
電話：（886）02-2237-1566　傳真：（886）02-2237-7135

2015年10月16日

加州（95014）庫比提諾無窮環路1號蘋果公司

參考：訂單編號333

致：傑克·斯諾先生

請供應以下貨品：

數量	描述	貨品編號	Unit Price
5	iPhone 6s (16GB) 金色	QA002	新臺幣$25,900
5	iPhone 6s (64GB) 金色	QA001	新臺幣$29,500
		總額：新臺幣$277,000	

付款期限應為，在10天內付款可享5%折扣，或在30天內付清全額。
請盡快以國際快捷寄送貨物。
請在2015年10月30日前運送所有貨品。

運送所有貨物到CCC公司

臺灣臺北市（106）大安區基隆路4段43號
電話：（886）02-2237-1566　傳真：（886）02-2237-7135

陳依菁

採購經理
CCC公司

APPLE
No. 1, Infinite Loop,
Cupertino, CA 95014

October 17th, 2015

CCC
No. 43, Sec. 4,
Keelung Rd.,
Da'an Dist., 106,
Taipei City, Taiwan R.O.C

Ref: Order Number #333

Attention: Ms. Yi Ching Chen

Thank you for your reply of 16th October. Our company is very glad to receive your order No. 333 for five iPhone 6s (16GB) Golden, item number QA002; and five iPhone 6s (64GB) Golden, item number QA001.

Special care will be devoted to the execution of your order. We will notify you as soon as possible when the consignment is ready for transport.

We hope this first order will lead to further business.

Sincerely yours,

Jacky Snow
Sales Manager
APPLE

中文翻譯

加州（95014）庫比提諾無窮環路1號　蘋果公司

2015年10月17日

臺灣（R.O.C）臺北市（106）大安區基隆路4段43號　CCC公司

參考：訂單編號333

致：陳依菁小姐

我們感謝您在10月16日的回覆。本公司很高興收到您的訂單：編號333，訂購5支iPhone 6s（16GB）金色手機，商品編號為QA002；以及5支iPhone 6s（64GB）金色手機，商品編號為QA001。

我們將致力為您的訂單執行專業支援。當貨品準備運送時，我們會盡快通知您。

我們希望這首次訂單能帶來以後更多的合作。

傑克・斯諾
銷售經理
蘋果公司

APPLE
No. 1, Infinite Loop,
Cupertino, CA 95014

October 23th, 2015

CCC
No. 43, Sec. 4,
Keelung Rd.,
Da'an Dist., 106,
Taipei City, Taiwan R.O.C

Dear Ms. Chen,

We want to say how pleased we were to receive your order of October 16th for iPhone 6s.

We would like to confirm the order of five iPhone 6s (16GB) Golden and five iPhone 6s (64GB) Golden at the prices stated in your Order No. 333. Our Sales Confirmation No. EF-555 in two originals was airmailed to you. Please sign and return one copy of them for our file.

It is understood that a letter of credit in our favor covering the said iPhones should be issued immediately. We want to point out that stipulation in the relative L/C must strictly confirm to those stated in our Sales Confirmation so as to avoid subsequent amendments. You may assure that we will deliver shipment without delay on receipt of your letter of credit.

We appreciate your cooperation and look forward to receiving further orders from you.

Sincerely yours,

Jacky Snow
Sales Manager
APPLE

中文翻譯

加州（95014）庫比提諾無窮環路1號　蘋果公司

2015年10月23日

臺灣（R.O.C）臺北市（106）大安區基隆路4段43號　CCC公司

親愛的陳小姐，

我們很高興收到您在10月16日訂購iPhone 6s手機的訂單。

我們確定以編號333訂單裡標示的價格，向您供應5支iPhone 6s（16GB）金色手機及5支iPhone 6s（64GB）金色手機。編號EF-555銷售確認書的兩份正本已航空郵寄給您。請簽署並寄回一份給我們做紀錄。

據理解，應立刻開立信用狀，內容須包含上述的iPhone手機規格。有關相關的信用狀規定，必須符合我們在銷售確認書裡所列明的，以免日後修改。您可以放心，我們收到信用狀後會立即出貨。

我們感謝您的配合，並期待以後收到您的訂單。

傑克・斯諾
銷售經理
蘋果公司

包裝
（正式商業書信寫作文體）

書信範例

CCC
No. 43, Sec. 4,
Keelung Rd.,
Da'an Dist., 106,
Taipei City, Taiwan R.O.C
Tel: (886)02-2237-1566 Fax: (886)02-2237-7135

October 30th, 2015

APPLE
No. 1, Infinite Loop,
Cupertino, CA 95014

Ref: Order Number #333

Dear Mr. Snow,

Thank you for your letter of October 28th informing us that iPhone 6s have been shipped by EMS and for your Invoice No. 314 in triplicate.

Regarding with the packing of iPhone 6s, you have made comments on packing the goods in boxes with blister packaging. After discussing the matter with our clients, we find that your suggestions are quite reasonable. Please take necessary precautions that the packing can protect the goods from dampness or rain.

You may assure that we shall do everything possible to work together. Thank you for your understanding.

Regards,

Yi Ching Chen
Purchasing Manager
CCC

中文翻譯

CCC 公司

臺灣臺北市（106）大安區基隆路4段43號

電話：（886）02-2237-1566　傳真：（886）02-2237-7135

2015年10月30日

加州（95014）庫比提諾無窮環路1號　蘋果公司

參考：訂單編號333

親愛的斯諾先生，

感謝您10月28日來信通知我們iPhone 6s手機已由國際快捷遞送以及一式三份編號314的請款單。

有關iPhone 6s手機的包裝，您提出以氣泡包裝盒包裝貨品，我們與客戶就這個問題討論後，覺得您的建議相當合理。請留意，包裝能防止產品受潮和接觸雨水。

我們雙方將盡一切努力充分合作。感謝您的諒解。

陳依菁
採購經理
CCC公司

CCC
No. 43, Sec. 4,
Keelung Rd.,
Da'an Dist., 106,
Taipei City, Taiwan R.O.C
Tel: (886)02-2237-1566 Fax: (886)02-2237-7135

November 20th, 2015

Fubon Insurance Company
No. 525, Zhongzheng Rd.,
Shilin Dist.,
Taipei City 111, Taiwan R.O.C

Dear Sir or Madam,

We are supplier of smart phones and electronic products in Taiwan. American Chamber of Commerce in Taiwan recommended your company to us.

We are interested in receiving the following information:

1. Insurance Types
2. Price Lists
3. Payment terms

Please send us the information requested above as soon as it is convenient for you. We are looking forward to receiving the information by November 25th. I appreciate your help.

Regards,

Yi Ching Chen
Purchasing Manager
CCC

中文翻譯

臺灣臺北市（106）大安區基隆路4段43號　CCC公司
電話：（886）02-2237-1566　傳真：（886）02-2237-7135

2015年11月20日

臺灣臺北市士林區（111）中山路525號富邦保險公司

親愛的先生女士，

我們是智慧型手機和電子產品的臺灣供應商，在臺灣的美國商會向我們推薦您。

我們有興趣收到以下的資訊：

1. 保險種類
2. 價目表
3. 付款細則

請在方便時盡快將以上要求的資訊發送給我們。我們期待在11月25日前收到回覆。感謝您的配合。

陳依菁
採購經理
CCC公司

書信範例

Fubon Insurance Company
No. 525,
Zhongzheng Rd.,
Shilin Dist., 111,
Taipei City, Taiwan R.O.C
TEL: (02) 2706-7890

November 22th, 2015

CCC
No. 43, Sec. 4,
Keelung Rd.,
Da'an Dist., 106,
Taipei City, Taiwan R.O.C

Dear Ms. Chen,

Thank you for your email of 20th November. We have received your inquiry about our insurance types, price lists, and payment terms.

We are pleased to provide you with this information in enclosure (1). To purchase any insurance, please either contact me at (02)-2706-7890 or send me an email.

Please don't hesitate to contact me if you need any further information. Thank you for giving us the opportunity to serve you.

Sincerely yours,

Tony Huang
Sales Manager
Fubon Insurance Company

Enclosure (1)

中文翻譯

臺灣臺北市士林區中山路525號　富邦保險公司
電話：（02）2706-7890

2015年11月22日

臺灣臺北市（106）大安區基隆路4段43號　CCC公司

親愛的陳女士，

感謝您在11月20日電郵來信。我們已收到您有關保險種類、價目表及付款細則的查詢。

我們很樂意向您提供附件裡的資料。如欲購買保險，請撥（02）2706-7890或再發電子郵件聯絡我。

如果您需要更多資訊，請馬上聯絡我。感謝你給予我們為您服務的機會。

黃湯尼
業務經理
富邦保險公司

附件(1)

書信範例

CCC
No. 43, Sec. 4,
Keelung Rd,
Da'an Dist., 106,
Taipei City, Taiwan, R.O.C
Tel: (886)02-2237-1566 Fax: (886)02-2237-7135

November 6th, 2015

APPLE
No. 1, Infinite Loop,
Cupertino, CA 95014

Ref: Order Number #333

Dear Jacky,

We would like to bring your attention to our Order No. 333 covering 10 pieces of iPhone 6s. We have sent you an irrevocable L/C-expiration dated on October 30th about 30 days ago.

When we placed the order we pointed out that the punctual shipment was of the utmost importance because this order was from the largest dealer here and we have given them a definite assurance that we could supply the goods by the end of October.

We would like to emphasize that any delay in delivery would undoubtedly involve difficulty.

Thank you very much for your cooperation.

Regards,

Yi Ching Chen
Purchasing Manager
CCC

中文翻譯

臺灣（R.O.C）臺北市（106）大安區基隆路4段43號　CCC公司
電話：（886）02-2237-1566　傳真：（886）02-2237-7135

2015年11月6日

加州（95014）庫比提諾無窮環路1號　蘋果公司

參考：訂單編號333

親愛的傑克，

我們希望您注意訂購10支iPhone 6s手機編號333的訂單，我們30天前已向你發送不可撤銷的信用狀，其限期至10月30日。

我們下訂單時指明準時發貨是最重要的，因為這次的訂購是由這裡最大的經銷商委託的，我們已向他們保證在10月底之前供應貨品。

我們想強調，出貨延遲無疑會帶來麻煩。

非常感謝您的配合。

陳依菁
採購經理
CCC

書信範例

APPLE
No. 1, Infinite Loop,
Cupertino, CA 95014

November 7th, 2015

CCC
No. 43, Sec 4,
Keelung Rd.,
Da'an Dist., 106,
Taipei City, Taiwan R.O.C

Ref: Order Number #333

Dear Yi Ching,

We are sorry to inform you that it has become impossible to complete shipment before the end of October of Order No. 333.

We have now in stock only a limited quantity of the items you required. It is impossible to ship within the validity of the Letter of Credit which expires on October 30th. Under the circumstances, we hope you will agree to extend the credit till November 30th as we request by email.

We also regret to notify you that an unexpected production problem on the part of our manufactures renders us unable to make immediate delivery of the goods. We are sorry and will do everything to expedite the manufacture. The expected date of the shipment will be around November 20th.

We hope you would understand the situation and will comply with our request.

Sincerely yours,

Jacky Snow
Sales Manager
APPLE

中文翻譯

加州（95014）庫比提諾無窮環路1號　蘋果公司

2015年11月7日

臺灣（R.O.C）臺北市（106）大安區基隆路4段43號　CCC公司

參考：訂單編號333

親愛的依菁，

很遺憾通知您，我們已無法在10月底前完成訂單編號333的送貨。

您需要的貨物，我們現在只有有限的存貨，所以不可能在信用狀的有效期（10月30日到期）內送貨。在這種情況下，我們希望您能同意如電子郵件裡所提及，把信用期延長到11月30日。

很遺憾地告知您，我們部份製造商出現意料之外的生產問題，以致我們無法即時發貨。我們對此同樣感到抱歉，會盡一切方法加快生產。預期大概11月20日能發貨。

希望您會體諒情況，並答應我們的請求。

傑克‧斯諾
銷售經理
蘋果

From: Yi Ching Chen [yichen@leeandsons.com.tw]
To: Customer Service [service@creditforall.com]
Subject: Incorrect Billing Charges

Dear Customer Service Representative,

I ordered twenty chairs from your company on November 27th. I received the right number of chairs, but it was charged for five more for a total of twenty-five chairs.

I don't know what went wrong with the bill. I would be grateful if you could check the bill again and give me the refund of five more chairs.

I look forward to your reply. In the meantime, if you have any question, please call or write to me.

Regards,

Yi Ching Chen
Leeandsons

中文翻譯

寄件人：陳依菁 [yichen@leeandsons.com.tw]
收件人：客戶服務 [Service@creditforall.com]
主旨： 錯誤的費用徵收

親愛的客戶服務代表，

我在11月27日在你們公司預訂了20張椅子。我收到了正確數量的椅子，但被多收取5張的費用——總共25張椅子。

我不知道帳單出了什麼錯誤。如果您可以再檢查一次帳單，並把那5張椅子的收費退還給我，我會很感謝。

我很期待您的回覆。在此期間，如果您有什麼疑問，請打電話或寫信給我。

陳依菁
李和森公司

書信範例

From: Customer Service [service@globalgarment.com.tw]
To: David Liang [david@privatecyber.com]
Subject: Re: Refund for Wrong Trousers (Order #C-9921A)

Dear David Liang,

We are sorry to hear that the color of the pair of trousers that you ordered was totally different from what you saw online. To make sure that our customers are satisfied with their items, we have a policy that allows to exchange products within a week of purchase.

The pair of trousers that you purchased from us is at least one month ago. We're sorry, Mr. Liang, but as the refund period has already expired, we cannot refund it. It is the first time that you have purchased from our online store, so we will provide you with an exchange of the item that you ordered online. Please place a new order on our online store and we will arrange for someone to pick up the wrong trousers from you.

We hope to serve you again in the future and look forward to seeing you at our online store soon.

Sincerely yours,

Jacky Snow
Customer Service Manager

中文翻譯

寄件人： 客戶服務 [service@globalgarment.com.tw]
收件人： 梁大衛 [david@privatecyber.com]
主旨： 關於：錯誤長褲的退費（訂單編號#C-9921A）

親愛的梁大衛，

我們很遺憾得知，您預訂的長褲顏色跟您在網路上看到的完全不同。為了確保顧客滿意他們所選購的商品，我們有一項政策，允許他們在購買一星期內退換商品。

您已購買這條長褲至少1個月。梁先生，我們對此感到很抱歉，但由於退款的期限早已結束，我們不能為您辦理退款。這是您首次在我們的網路商店購物，因此我們向您提供商品替換，更換您之前所訂購的。請在我們的網路商店重新下訂，我們將安排專人到您府上回收錯誤的長褲。

我們希望以後能繼續為您服務，並期待再次在網路商店見到您。

傑克・斯諾
客戶服務經理

From: WeiHan Wang [weihan@gmail.com]
To: Sally Lin [Sallyl@gmail.com]
Subject: Problem About Book Order

Dear Sally,

I have ordered 15 different books from your company on August 1st, 2016. I have told you that the books are needed by August 20th, and you have assured that there would be no problem.

When I received the books, I found out that your company had sent me the wrong number of books—only eleven out of fifteen. You promised to take care of it, but this time, you sent me the wrong books—the four I already received the first time!

When I called up and asked Mark Sung to explain the problem, he was rude and uncooperative. He told me "you had no reason to blame us".

I am disappointed with the way that your company dealt with the problem. Mark Sung's behavior is totally unacceptable, and I refuse to let him handle this order any longer. Please assign another Sales Manager to deal with this order. Please send me the four books I haven't received as soon as possible by August 20th. I hope this problem will be taken care of immediately.

Regards,

Wei Han Wang

中文翻譯

寄件人：王蔚涵 [weihan@gmail.com]
收件人：林莎莉 [Sallyl@gmail.com]
主旨：　有關書籍訂購問題

親愛的莎莉，

我已在2016年8月1日訂購15本不同的書，並已經告知我需要在8月20日前拿到，而貴公司保證不會出錯。

我收到這些書時，發現貴公司送錯書的數量了，15本裡只有11本。你答應會處理好這件事，但這次，發送了錯誤的書，有4本是我在第一次已收到的！

當我打電話要求宋馬克解釋這個問題，他沒有禮貌而且不肯合作。他跟我說：「你沒有理由責怪我們。」

我對你們公司解決問題的方式很失望。宋馬克的態度讓人完全沒法接受，我不會再允許他處理我的訂單。請調派另一位銷售經理處理這次訂購。請把剩下的4本書盡快在8月20日前寄給我。我希望這個問題能馬上獲得關注。

王蔚涵

書信範例

From: Sally Lin [Sallyl@gmail.com]
To: WeiHan Wang [weihan@gmail.com]
Subject: Re: Problem About Book Order

Dear Ms. Wang,

Thank you for letting me know about Mark's behavior. We are sorry to hear that you have not received the right books and we apologize for any inconvenience caused.

We will assign another Sales Manager, Daniel to deal with your order, and he will contact you as soon as possible. I have also scheduled a meeting to discuss with Mark's attitude and bad behavior about this order.

I will also personally take care of this order from here until you receive the four books. It would be a shame for us to lose a valuable customer like you, and I hope that the problem will be solved immediately.

Thank you again for getting in touch with me directly.

Warm Regards,

Sally Lin
Sales Director

中文翻譯

寄件人： 林莎莉 [Sallyl@gmail.com]
收件人： 王蔚涵 [weihan@gmail.com]
主旨： 回覆：有關書籍訂購問題

親愛的王小姐，

感謝您讓我知道有關馬克的表現。我很抱歉得知您沒有收到正確的書籍，並為您帶來不便。

我們會調派另一位銷售經理處理您的訂單，他會盡快聯絡您。我還安排了一個會議，討論馬克的態度和針對這次訂購的惡劣表現。

我也會親自處理這次訂購，直到您收到那4本書。失去像您這麼寶貴的顧客會是我們的恥辱，我希望這件事會馬上解決。

再次感謝您直接聯絡我。

林莎莉
業務處長

書信範例

CCC
No. 43, Sec. 4,
Keelung Rd.,
Da'an Dist., 106,
Taipei City, Taiwan R.O.C
Tel: (886) 02-2237-1566 Fax: (886) 02-2237-7135

December 11, 2015

APPLE
No. 1, Infinite Loop,
Cupertino, CA 95014

Ref: Order Number #333

Dear Mr. Snow,

We enclosed an order for 5 pieces of iPhone 6s (16GB) & 5 pieces of iPhone 6s (64GB) and have made an arrangement with Citibank, Taipei, to issue the L/C in your favour. The L/C is valid until December 31st and will be confirmed to you by the Bank's California office. Please draw on them at 30 d/s. It is understood that the goods will be shipped by S.S. Kitty sailing from California on December 24th.

Cases containing the goods should be marked C.C.C. The amount of our credit has been fixed to provide adequate cover for your invoice, which will provide all charges, including insurance to Taiwan.

Please notify us when the goods are shipped.

Regards,

Yi Ching Chen
Purchasing Manager
CCC

Enclosure 1

中文翻譯

臺灣（R.O.C）臺北市（106）大安區基隆路4段43號　CCC公司
電話：（886）02-2237-1566　傳真：（886）02-2237-7135

2015年12月11日

加州（95014）庫比提諾無窮環路1號　蘋果公司

參考：訂單編號333

親愛的斯諾先生，

我們附上5張iPhone 6s（16GB）和5張iPhone 6s（64GB）的訂單，並已向臺北花旗銀行做出安排，為您開立信用狀。信用狀期限至12月31日，會由您在加州的銀行辦公室確認。請在見票後30天動用。據理解，貨物會在12月24日經由S.S凱蒂航海從加州運送。

貨品的外盒應印有C.C.C標誌。我們的信用金額已定，足以涵蓋請款單的全部金額及貨物運送臺灣的保險。

貨物發送時，請通知我們。

陳依菁
採購經理
CCC

附件1

出貨通知
（正式商業書信寫作文體）

APPLE
No. 1, Infinite Loop,
Cupertino, CA 95014

December 14th, 2015

CCC
No. 43, Sec. 4,
Keelung Rd.,
Da'an Dist., 106,
Taipei City, Taiwan R.O.C

Ref: Order Number #333

Dear Yi Ching,

Your order No. 333 of October 16th has now been completed and loaded on board. S.S. Kitty will be sailing tomorrow.

Shipping documents, including original invoice, have been sent to the California Branch of Citibank, upon whom we have drawn at 30 d/s for the net amount due, NTD$ 300,000.

We trust the consignment will reach you safely and satisfactory. We appreciate your cooperation and look forward to receiving further orders from you.

Sincerely yours,

Jacky Snow
Sales Manager
APPLE

中文翻譯

加州（95014）庫比提諾無窮環路1號　蘋果公司

2015年12月14日

臺灣（R.O.C）臺北市（106）大安區基隆路4段43號　CCC公司

參考：訂單編號333

親愛的依菁，

您在10月16日的訂單，編號333，已處理完成，S.S.凱蒂將會在明天啟航。

送貨文件，包括單據正本已遞送到花旗銀行加州分行。其中，我們在見票後30天動用應有的淨金額，新臺幣$300,000元。

我們相信貨物會安全送到您手上，並讓您滿意。我們感謝您的配合，期待以後再收到您的訂單。

傑克·斯諾
銷售經理
蘋果公司

From:　　Yi Ching Chen [Yi Ching@gmail.com]
To:　　　Janet Fan [janetfan@gmail.com]
Subject:　Congratulations on New Business!

Dear Janet,

Congratulations on the opening of your new branch office.

Your new office, conveniently locating in the financial district, is beautifully decorated and well-lit. It offers a nice environment to your colleagues. There are several recreation rooms in your new branch office where your co-workers can release their stress and relax themselves. In addition, the cafeteria serves various kinds of fresh and delicious cuisines. I believe all the colleagues will be full of energy and enjoy work life.

I wish you success with your new business venture. If there is anything I can do to make it easier for you to get started in your new business, please let me know.

Best wishes,

Yi Ching Chen

中文翻譯

寄件人：陳依菁 [yichen@leeandsons.com.tw]
收件人：范珍妮特 [Service@creditforall.com]
主旨：　祝賀您的新業務！

親愛的珍妮特，

恭喜您新分公司開張。

您的新公司，位於交通便利的金融區，建設漂亮且光線充足。我覺得能為您的員工提供一個美好的工作環境。新分公司有多個娛樂室，可以讓您的同事釋放壓力、放鬆自己。此外，公司裡的自助餐館供應各種新鮮美味的菜色。我相信，所有員工都會充滿能量並享受工作。

我希望您的新企業成功。如果有任何地方我可以幫助您，讓您更容易開展新業務，請讓我知道。

陳依菁

回覆恭賀信件
（非正式電子郵件）

From: Janet Fan [janetfan@gmail.com]
To: Yi Ching Chen [Yi Ching@gmail.com]
Subject: Re: Congratulations on New Business

Dear Yi Ching,

Thanks for your kind words. What's more, thank you for coming the opening of our new branch office. I am glad that you like the facilities and the decoration in our new branch office. We have spent a lot of time and efforts on them. We will manage and maintain them well in the future.

I can not hide my joy, I must say. I'm truly happy to be the general manager of this new branch office. Now I hope to make this new business successful.

Thanks for letting me know that you will be there, in case I need help. I would write an email to tell you my latest news.

Sincerely yours,

Janet Fan

中文翻譯

寄件人：范珍妮特 [janetfan@gmail.com]
收件人：陳依菁 [Yi Ching@gmail.com]
主旨：　回覆：祝賀您的新業務！

親愛的依菁，

感謝您那麼客氣。更重要的是，謝謝您光臨我們新分公司的開幕式。我很高興您喜歡新辦公室的設施和裝潢。我們在這些事上花了很多時間和努力。未來我們會好好管理和維持它們。

我必須說，我無法掩飾我的喜悅。我真的很高興成為這家新分公司的總經理。我現在的願望就是努力讓企業成功。

感謝您讓我知道萬一我需要幫助，您會在身邊支持我。我會寫信告訴您我最新的消息。

范・珍妮特

國家圖書館出版品預行編目（CIP）資料

商用英語立可貼／林奇臻著.
--初版, -- 臺北市：大是文化，2016. 09
　　　面；　　公分.（Biz；203）
ISBN 978-986-5612-71-9（平裝）
1. 商業書信　2. 商業英文　3. 商業應用文

493.6　　　　　　　　　　　105013088

Biz 203

商用英語立可貼

上網開店、網拍、揪團購、甚至開公司，報價、詢價、殺價、訂約收錢，
就算菜英文，複製貼上就搞定

作　　　　者／林奇臻
責 任 編 輯／劉宗德
校 對 編 輯／王怡婷、廖恒偉
美 術 編 輯／邱筑萱
副 總 編 輯／顏惠君
總　　編　　輯／吳依瑋
發　　行　　人／徐仲秋
會　　　　計／林妙燕
版 權 主 任／林螢瑄
版 權 經 理／郝麗珍
行 銷 企 畫／汪家緯
業 務 助 理／馬絮盈、林芝蓉
業 務 專 員／陳建昌
業 務 經 理／林裕安
總 經 理／陳絜吾

出　　　　版／大是文化有限公司
　　　　　　　台北市100衡陽路7號8樓
　　　　　　　編輯部電話：（02）2375-7911
讀 者 服 務／購書相關資訊請洽：（02）2375-7911　分機122
　　　　　　　24小時讀者服務傳真：（02）2375-6999
　　　　　　　讀者服務E-mail：haom@ms28.hinet.net
郵政劃撥帳號／19983366　　戶名：大是文化有限公司

香 港 發 行／里人文化事業有限公司 "Anyone Cultural Enterprise Ltd"
　　　　　　　香港新界荃灣橫龍街78號　正好工業大廈22樓A室
　　　　　　　22/F Block A, Jing Ho Industrial Building, 78 Wang Lung Street, Tsuen Wan,
　　　　　　　N.T., H.K.
　　　　　　　電話：852-24192288　　傳真：852-24191887

封 面 設 計／林雯瑛
內 頁 排 版／黃淑華
印　　　　刷／鴻霖印刷傳媒股份有限公司

■ 2016年9月初版　　　　　　　　　　　　　　Printed in Taiwan
ISBN 978-986-5612-71-9（平裝）　　　　　　　定價／新台幣350元
　　　　　　　　　　　　　　　　　　　　（缺頁或裝訂錯誤的書，請寄回更換）